DAN

Thank you for
all your efforts to
enlighten the world through
The L.G. magazine. I
appreciate your writings and
insightful words

Tom Chattan
6/10/2023

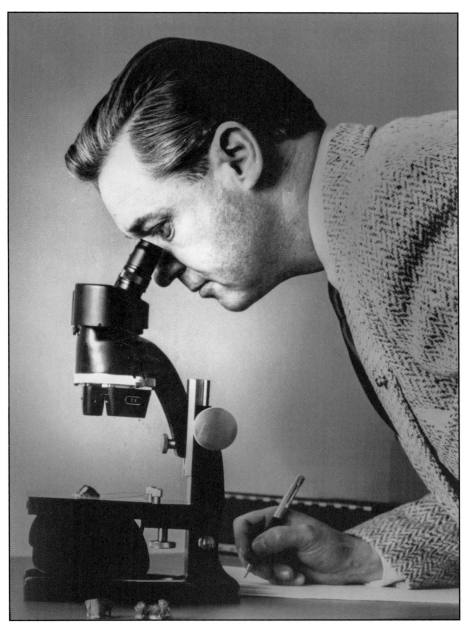
Carroll Chatham spent most of his life looking through a microscope, tinkering in his lab and fighting to get his gemstones into the hands of people who love jewelry.

The Chatham Legacy: An American Story
By Thomas H. Chatham

Editor:
Amanda J. Luke

Design:
Orasa Weldon

Cover: Carroll Chatham in his first lab using glass apparatus he made by hand because what he needed was not available in local shops.

Back cover: Carroll (left) and Tom Chatham in the shop area of the lab working on a problem.

Left: Chatham Created ruby.
Credit: Orasa Weldon.

Thomas H. Chatham
Chatham Created Gems & Diamonds, Inc.
300 Rancheros Dr.
San Marcos, CA 9206
www.chatham.com

Library of Congress Control Number: 2023905715

Printed in the United States of America

Copyright © 2023 Thomas H. Chatham

All rights reserved. No part of this publication may be reproduced, distributed, or transmitted in any form or by any means, including photocopying, recording, or other electronic or mechanical methods, without the prior written permission of the publisher, except in the case of brief quotations embodied in critical reviews and certain other noncommercial uses permitted by copyright law.

To request permissions or to order copies, contact the publisher at sales@chatham.com.

Hardcover: ISBN 979-8-9879161-0-0

Printed by Neyenesch Printers Inc.
2750 Kettner Blvd.
San Diego, CA 92101
www.neyenesch.com

Table of Contents

Introduction_x

Dedication_xii

Foreword_xiii

1. Entrepreneurial Roots_1
2. Genius in the Bud of Life_5
3. The First Chatham Laboratory-Grown Emerald_12
4. The Birth of Chatham Research Laboratories_16
5. Guerilla Marketing, Chatham-Style_24
6. The Demise of the Loop Lumber & Mill Company_32
7. A Swiss Partner Swoops In_36
8. The Backlash of the Industry, Courtesy of the FTC_39
9. Whoops ... the FTC Changed its Mind!_46
10. Lessons Learned Growing Up_52
11. Inventing an Emerald-Growing Furnace_56
12. How to Grow Emeralds_62
13. Upsetting the Profit-Sharing Apple Cart_68
14. Acceptance in the Trade_73
15. Staying Ahead of the Competition_80
16. Navigating Possible Big Buyouts_86
17. The Difficulty of Being Independent_92

18. Doing Business in 1970s Hong Kong_96

19. Oswald Dallas and His Cutting Factory_100

20. Meanwhile, Back in the USA_105

21. African Adventures _109

22. Dealing with the Cartel in Colombia_116

23. Finally, Chatham Created Gems, Inc. is Independent_124

24. The Emerald Wars_129

25. Gem Trade Lawsuits_133

26. Re-evaluating Our Next Moves_138

27. Overcoming Disasters_144

28. Diamond: The Holy Grail of Crystal Growth_150

29. Seeking Business Partners in Russia_154

30. Trying to Do Business in Ukraine_162

31. Introducing the Chatham Created Diamond_170

32. The Partnership of Tracy Hall and Chatham_178

33. The Rebirth of the Japan Connection_184

34. The Evolution of Chatham Created Gems & Diamonds_190

35. Marketing with Integrity_196

36. How to Travel Safe and Work the Trade Shows_200

37. The Future of Chatham_214

Afterword_220

Acknowledgements_222

Further Reading_223

About the Author_230

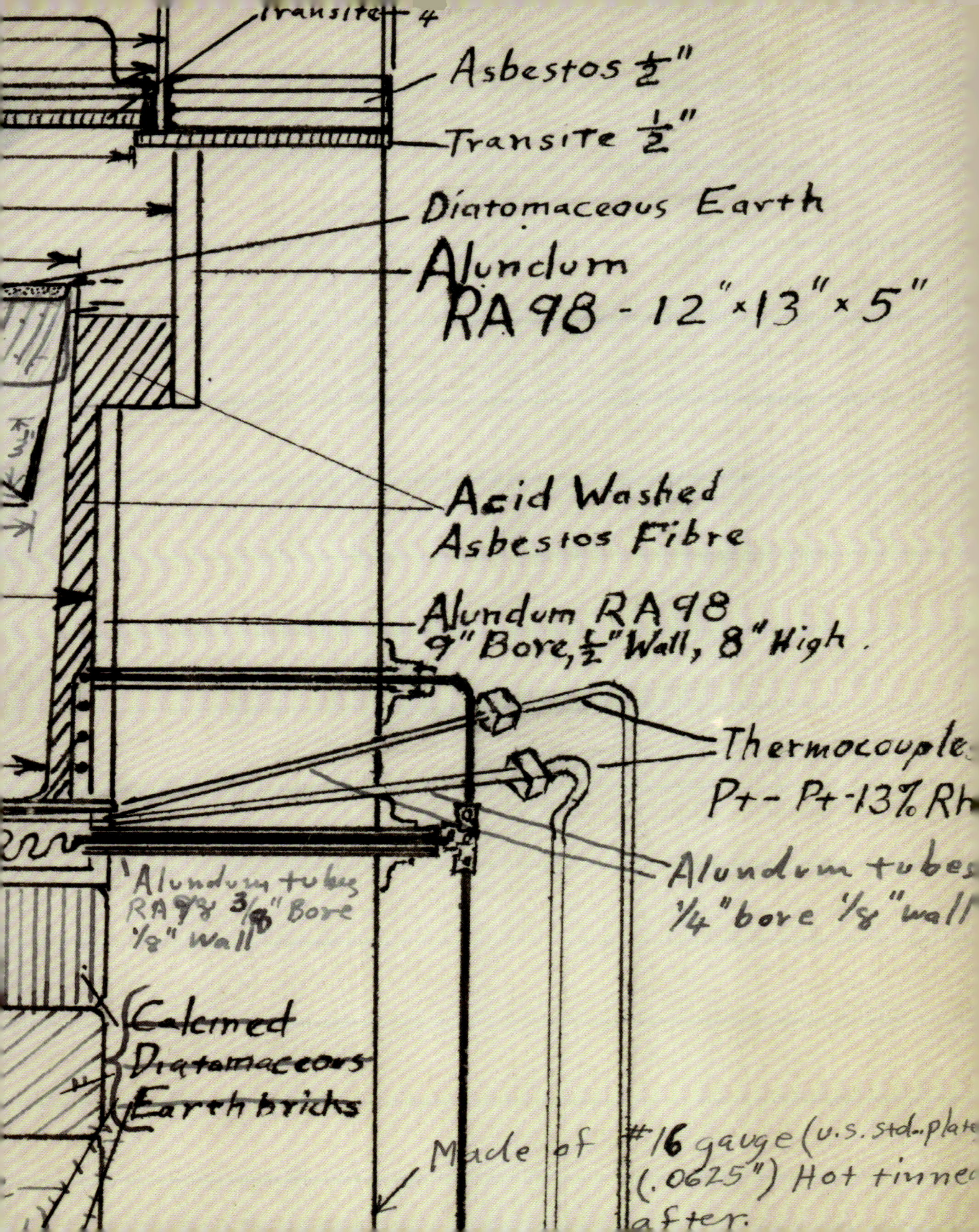

Transite 4

Asbestos $\frac{1}{2}$"

Transite $\frac{1}{2}$"

Diatomaceous Earth

Alundum
RA 98 - 12" x 13" x 5"

Acid Washed
Asbestos Fibre

Alundum RA 98
9" Bore, $\frac{1}{2}$" Wall, 8" High.

Thermocouples
Pt - Pt - 13% Rh

Alundum tubes
$\frac{1}{4}$" bore $\frac{1}{8}$" wall

Alundum tubes
RA 98 $\frac{3}{8}$" Bore
$\frac{1}{8}$" Wall

Calcined
Diatomaceous
Earth bricks

Made of #16 gauge (u.s. std. plate)
(.0625") Hot tinned
after.

Carroll Chatham was featured in "The man who Grows Emeralds," a *True* magazine article by J.P. Cahn in 1957. Credit: *True* magazine.

Introduction

In the development of any area of human endeavor, there are always a few specific individuals who, through a combination of intelligence, commitment, persistent effort and some luck, make significant contributions to that undertaking.

In the field of gemology, one such individual was Carroll Chatham (1914-1983), an American chemist who developed an interest in laboratory-grown gems at an early age. Following several years of experimentation on crystal growth in his home and combined with a college chemistry degree from the California Institute of Technology, he was finally able to grow one-carat emerald crystals using a flux solution method in 1935, and then on a reproducible basis in 1938.

In this volume, Carroll's son Tom presents for the first time the fascinating history of his father's creation of emerald followed by other laboratory-grown gem materials. He also describes their lengthy efforts to bring this material into the gem market. Some of this information has not been previously published or presented in this amount of detail.

Growing crystals in the laboratory is both a science (follow the correct process) and an art (learn from your successes and mistakes), and both aspects are discussed here.

Emeralds were first grown in a lab in France by the flux method in 1888, but although of good color, the crystals were too small (~ 1 mm) to be of any practical use. Laboratory-grown ruby and sapphire was created by a melt technique (the flame fusion or Verneuil method) in the early 1900s. For several decades, however, there were few efforts to produce laboratory-grown emerald crystals on an industrial scale until the successful results obtained by Carroll Chatham.

This book chronicles the years of work that were needed to first grow emeralds in a laboratory, and then the trial-and-error experiments and design modifications developed to expand and optimize the growth process on a larger scale. This was followed by the initial efforts to sell this new material into the jewelry trade, a technically conservative industry whose members are often hesitant to adopt new products – especially those that were claimed to have been made in a laboratory. Proper identification of these laboratory-grown gems (as opposed to natural emeralds) was also a challenge for jewelers with little or no gemological training and equipment.

An important part of the book describes Chatham's expensive and years-long legal disputes with the Jewelers Vigilance Committee and the Federal Trade Commission (FTC) over this material being called a synthetic gem. Carroll Chatham strongly disagreed with this designation and the dispute was finally resolved in 1964 when the FTC decided that the marketing name "Chatham Created Emerald" was acceptable. This effort was followed by attempts by the company to be allowed to participate as a member in major jewelry trade organizations and trade shows, and to sell their production in rough or polished form directly to buyers.

This book represents one of the rare firsthand accounts of the creation and marketing of a new line of laboratory-grown gem materials, with the author being personally involved with many of the steps of the process. It is also a story about the close relationship between a father and his sons and how each individual contributed the different technical, business and marketing skills required to bring such a venture to fruition. Their success was achieved through imagination (in the lab and in the gem and jewelry industry), determination, cooperation and respect for each other's strengths.

Dr. James Shigley

Distinguished Research Fellow at the Gemological Institute of America.

Dedication

This book is dedicated to the memory of my father, Carroll Chatham, an outstanding chemist and a talented fabricator. He taught by example and was rarely disappointed in his mistakes or excited by his accomplishments.

He was a thinker – a plodder who persevered in everything he attempted to do – and he was quiet as he worked his way through possible outcomes and scenarios. His mind was like a Google search engine that memorized the handbook of inorganic chemistry, a formidable task. He was proficient in German because many chemistry books were only written in German. He was also proficient in glass blowing because he needed to be able to create his own glass vessels for his experiments; the type he needed were not available in stores.

As a father, he was usually missing and could be found in his laboratory. As a boss, he was instrumental in making me what I am today. He was patient with those who didn't understand chemistry. He taught me to not repeat past mistakes in crystal growth research and to always seek answers in chemistry, no matter how difficult it may seem. I was often guilty of the "when in doubt, throw it out" type of problem solving, but not Carroll. He taught me to think and analyze results and gave me the confidence to go forward with unproven concepts while exhibiting unbelievable patience. He encouraged me to be better than I thought I was.

There are very few Carroll Chatham's in this world, and I feel privileged to share his limelight.

Foreword: The World of Chatham

This is a story about my family's contributions to the gem and jewelry industry. It is an autobiographical trail, from the mid-1800s to present times, that encompasses three generations of the Chatham family. Some dates are fuzzy, but most are close to reality.

Why the focus on three generations and not just one? Because each contributed either directly or indirectly to the legend of Carroll Chatham and his emerald discovery. Each was, or is, responsible for creating the legacy the Chatham name holds today.

My father's curiosity and intelligence led him to produce the first laboratory-created emeralds on a commercial basis. This accomplishment was followed by the creation of the first flux-grown ruby; flux-grown alexandrite; blue sapphires, padparadscha, pink sapphires, yellow sapphires; opal; and eventually, a flux-grown, high-pressure, high-temperature created diamond. I joined him in his efforts in my early 20s and helped him build his scientific passion into an international created gemstone business. Before Carroll Chatham, there were only a few laboratory curiosities of man-made emerald, but none that had any commercial value. I don't want to diminish the work of these pioneers, but what they did came nowhere close to what my father accomplished.

Nomenclature is a touchy subject in the gem business. Ask a natural stone dealer what Chatham Created Gems, Inc. produces, and he may say "synthetics." No love lost there. We, in turn, and all other crystal growers, will not – nor are required – to use the word "synthetic" to describe our products. Even the Federal Trade Commission has boldly stepped forward and advised against its use to avoid confusion in the marketplace.

A consumer thinks "synthetic" means fake and in many products their conclusions are correct. Synthetic rubber is not identical to natural rubber; nor is nylon or many plastics that resemble a natural product. Synthetic oil did not come from the results of bio fermentation over billions of years. The list can go on and on.

So instead, laboratory-created – or Chatham Created – is the appropriate moniker. In this book about how Chatham Created Gems, Inc. began, I am only discussing laboratory-grown gemstones and in the interest of brevity, have taken the liberty of dropping "created" in every instance.

This is a story about challenges: how to create the best processes and environments to grow gems; acceptance of them in the gem and jewelry trade – including courtroom dramas with the Federal Trade Commission; setting up business partnerships around the world; and finally, getting the word out about our products.

It is a story filled with adventures, people and a multi-generational passion for beautiful gemstones.

1

Entrepreneurial Roots

It was an exciting and intoxicating time, long after the gold rush of 1849 … [My grandfather] William ran a big, profitable lumber business that took in about $3 million a year; lumber was in short supply and prices were sky high.

Carroll Chatham loved chemistry, anything chemistry. He was an avid reader of science books and loved to recreate experiments. Luckily, when he was a young boy, chemicals were readily available at the local drug stores, then called the Chemist Shop. If you knew what to ask for, you could have a lot of fun.

So, Carroll made a lab in his family's cellar, with a window right at sidewalk level and behind the protection of a concrete foundation. He taught himself how to make rockets and explosives with ingredients he got at the Chemist Shop. Many were ignited outside this same window, regardless of who might be walking down the street.

As a teenager, he decided to recreate – and modify – an experiment he read on how to make a diamond. The resulting explosion took out the windows in the house across the street and completely enveloped his house in nitrogen gas, which luckily was inert and not poisonous. As for any pedestrians out walking their little poodles down the sidewalk? Fortunately, there were none.

There were plenty of police, however, and the results of the experiment were never known. The ingredients were totally obliterated and spread throughout the street. Maybe he did make diamonds, but there was no evidence of them to be found in the neighborhood.

What he did find was trouble and a reprimand from his father, William, who had no idea what his son was doing in his makeshift basement laboratory. His word was final: No more experiments and no more explosions that brought both neighbors and the police down on his household.

Carroll got the message, loud and clear, but never lost his desire to make diamonds.

William Chatham's house, at about 8,000 square feet and 3-stories high, is a fitting example of the craftsmanship of the time and it still stands off 37th Avenue near Geary Street. My grandfather's 1926 Cadillac is parked in front of the house in this photo. The small windows on the bottom right are where my father Carroll Chatham's very first lab was located when he was a young boy.

William Chatham, Carroll's father and my grandfather, was born in 1862 in San Francisco to Roland and Diana Chatham. The family name is of English origin, town unknown. All my searches have turned up little except that William was a seafaring captain who piloted a sizeable schooner up and down the West Coast transporting lumber.

It was an exciting and intoxicating time, long after the gold rush of 1849. Wood was in high demand and fortunes were made after the 1906 earthquake and firestorm that almost leveled San Francisco. The Loop Lumber & Mill Company, based in Alameda and San Francisco, California, had five ships, some towing raw lumber down the West Coast or shipping finished lumber all over the world.

One of the Loop Lumber & Mill Company's ships was named the F.S. Loop which was 193 feet long and weighed 790 tons. Another was the Mable Gale and two others of unknown name.

My grandfather's ship was registered as the SS William Chatham and guided raw cut tree trunks down to Alameda to be milled into building material or telephone poles and then transported to the San Francisco yard at the foot of 16th St. Somewhere along the way, William Chatham became the owner of the Loop Lumber & Mill Company, which he incorporated in 1925 as the sole owner.

William ran a big, profitable lumber business that took in about $3 million a year; lumber was in short supply and prices were sky high. He hired hundreds of men from

The April 19, 1906 front page of *The New York Times* with news of the San Francisco earthquake. Credit: *The New York Times*.

Asia to South America to work on the docks of San Francisco, a major port of destination and embarkation. No doubt the men were underpaid and overworked, but one needed a job to eat. Thousands of Chinese men had come to America to help build the railroads and work in the gold fields. They were free but worked cheap. Dirt cheap.

That began to change around 1920 with the arrival of Harry Bridges, a young man from Australia who had grand ideas to organize labor. He would make proclamations such as "Unite! Stand up to the owners and rich politicians and get a fair day's pay!" Bridges tried to organize everyone who worked offloading the ships and filling up warehouses, basically the ones who did all the hard labor in and around the bay's loading docks.

Bridges once called on William Chatham in his upstairs office at the foot of 16th St. to see if he could talk to Loop Lumber's employees. "Why?" asked William, knowing damn well why. "We want the employees to be represented by fair-minded people, like me," Bridges said. Chatham threw Bridges down the stairs to the street and told him "Don't ever step foot on this property again."

Bridges eventually established the International Longshoreman's and Warehouse Union (ILWU) in 1937, and it became one of the most powerful unions in the country. It was also the beginning of the end for Loop Lumber's business.

Before that happened, however, William decided to build a house out in the sand dunes near the western edge of San Francisco in the early 1920s. The area later became known as Sea Cliff.

It was a magnificent home that cost the staggering sum of $14,000 to build in 1920. Today, its value is more than $5 million.

William and his wife Madeline had four children: William Jr., Russell, Ruth and Carroll. Ruth, close in age to Carroll, had a lot to do with his upbringing when their mom died at the young age of 61. Madeline lived in the house for 16 years before her death in 1936. The four children were raised extravagantly but were not spoiled. Grandpa Chatham was a stern taskmaster and made sure everyone appreciated what he had created. The house was run by several maids, cooks and maybe a nanny, all who lived on the property.

{3}

The Loop Lumber & Mill Company supplied the lumber to build the 6-story Hotel Piedmont in 1928. Credit: *Oakland Tribune*, May 20, 1928.

I remember visiting the home for Christmas dinner in 1950 with all my aunts, uncles and cousins. I was about 5 years old, dressed in my little shorts with suspenders for a very formal meal. Everyone sat around one table that seemed like it was 40 feet long – I had never seen so many glasses, silverware and plates!

Grandpa Chatham picked up a little porcelain bell and shook it three times. Voila! Out from the kitchen paraded four well-dressed Asian servants, each with a white cap and a long-braided ponytail, carrying trays of food. We did not live this way in my father's household, and I was mesmerized and overwhelmed. I thought to myself then, and I remember it like yesterday, "When I grow up, I want one of those bells!"

After dinner, the kids romped around the house, played with the elevator and electric chair on the stairs and went down into the old and moldy smelling basement. I remember lifting the lid of an ornate wooden box and saw my first hand-cranked phonograph player, though at the time, I had no idea what it was. Then I went into a room that was about 20 by 12 feet that smelled of chemicals and I held my nose. All along the walls were bottles and bottles of chemicals that we were warned not to touch. A human skull with a hat on it and a pipe in its mouth sat on top of a wooden desk.

That Christmas was one of the few times I saw my Grandpa Chatham's wealth and the world he lived in.

2
A Genius in the Bud of Life

Carroll was always on the lookout for problems to solve and often went to the police crime lab to ask questions. He learned that they were having trouble taking pictures at night (flash bulbs were not invented yet), so he began work on a flash powder that would provide light for the camera.

Carroll, far left, at about 12 years old, and his friends Tom and John in 1926. The boys were testing an experiment with flash powder to see if it would provide enough lighting for the camera for night photography. They weren't concerned with posing for the photo, they just wanted to know if they could be seen – and they were successful.

Carroll F. Chatham was born on Sept. 9, 1914. His father was 53 years old at his birth, and his mother was 40, an unusual circumstance back in those days, although Carroll was the last of four children and was perhaps also a bit unexpected. Upon Madeline's hospital release, William was informed of the requirement of a middle name for his newborn son Carroll, so he said "F, just F." Years later I asked my father what the "F" stood for, and he said, "It doesn't stand for anything, but maybe my father was making a joke of the situation and used 'F' as in FINAL!"

Carroll attended Lafayette Elementary School, where he must have impressed someone, because when I went to the same elementary school 30 years later, a few of his old teachers were still there and commented, "I hope you are as smart as your father was!"

Carroll's love of chemistry and experiments began at a very young age, and he built a lab in the basement of the family home. He took advantage of the readily available ingredients at the Chemist Shop.

Today, just being in possession of these chemicals is against the law. When people abused the available ingredients and were hurt or killed, the Chemist Shops evolved into what we now know as drug stores and don't sell raw chemicals.

Carroll was always on the lookout for problems to solve and often went to the police crime lab to ask questions. He learned that they were having trouble taking pictures at night (flash bulbs were not invented yet), so he began work on a flash powder that would provide light for the camera. Flash powder is just gunpowder "slowed down" in reaction – pure gunpowder "booms" too quickly and destroys the camera.

Carroll and his two neighborhood sidekicks, Tom McGlynn and John (who became a very successful physician, but whose last name I can't recall), helped him in his endeavors. Once he thought the flash powder was successfully made, they tried it out on his front lawn with a camera set up. The camera was electrically slaved to the flash powder container so that when the camera was clicked (somehow remotely or by another buddy) the powder flashed and illuminated the yard. I don't know if the crime lab used his concoction, because after all, he was only a teenager, but it did work!

Another remarkable experiment my father attempted in his early years was the recreation of the Dr. Henri Moissan diamond experiment of 1865.

The components of diamond were discovered in the late 1800s, when scientists burned a diamond and analyzed the gases coming off it. They found that the diamond was 100 percent carbon dioxide. The carbon was like graphite, but in a different crystalline form. Thus, began a 200-year quest to recreate diamond. Robert Hazen's, "The Diamond Makers," is an excellent read on the topic.

Moissan speculated that diamond must form deep in the Earth, where pressures are immense and the temperatures very hot. By analyzing the materials found in the host rock of diamond mines, it was determined that iron and nickel were also present. Moissan theorized: What if one was to heat a mixture of carbon, iron and nickel to a melt, then somehow create pressure – would diamonds form?

This hypothesis was easier said than done. First, no furnaces at the time were hot enough to reach a melting point of nearly 1,200 Celsius, which Moissan determined might not be hot enough, so he invented the electric arc furnace. He had his "melt" method, but where was the pressure going to come from? He figured that when something very hot in liquid form hits another liquid in a much cooler form, the hotter substance will harden and contract, creating the necessary pressure.

Moissan used molten lead as his "cooler" liquid and his carbon-iron-nickel melt at 2,000 Celsius as the hotter liquid. He poured the hotter liquid into the cooler for his experiment. His theory of the mechanics involved was correct and a ball of iron was extracted from the lead. When the ball was cooled and split open, he discovered that tiny white crystals had formed inside. Moissan dug out a few of the white colorless crystals and used them to scratch various known materials he had at his disposal. When he reached corundum, the second hardest thing known to man at the time, the crystal created a scratch on it, so he declared success: diamond was created!

Moissan published papers on his experiment and presented them at various scientific lectures in Europe and the United States. Keep in mind, this was the late 1800s and sophisticated testing methods were not yet invented.

As technology and testing equipment progressed over time, other scientists determined that Moissan had not made diamond, but a new material: Moissanite, which was named in his honor. Years later, a natural form of the material was found, and the experts had a dilemma: man-made materials with a natural counterpart do not get the "ite" designation so they tried to change it but obviously failed. The man-made version is still called Moissanite, and the natural version is called silicon carbide, which is harder than corundum, but softer than diamond.

Moissan later went on to identify and isolate a new element, fluorine, which unfortunately is extremely toxic. His instrumental discovery of element number 9 and his continued research eventually killed him.

Carroll read all about Moissan's experiment and decided to repeat it, but with a twist. He didn't believe in repeating other's mistakes, which only led to the same results. This was a practice he lived by his entire life: If it doesn't work the first time, don't do it again. Go a different direction, add a little of this, just do something a little different. This has repeatedly proven to be of great value in crystal growth. Some of the simplest of changes – some by mistake – can prove invaluable.

The premise of carbon, iron and nickel made sense in the previous experiments and so did the pressure, but Carroll was not sure how to increase the pressure. He decided to also heat the iron, nickel and graphite to 2,500 Celsius, but drop it into liquid nitrogen at a minus 200 Celsius, which was well below the melting point of the lead used in Moissan's experiment. He put a container of the liquid nitrogen on the sidewalk through the little window of his basement laboratory and rigged the container of the melted iron and carbon mixture with a string so he could "safely" dump it into the nitrogen.

What a reaction! Both containers and the contents were obliterated, the police arrived, and Carroll's father put an end to explosive experiments in or by his basement.

Carroll was enrolled in Lowell High School, located many miles from home, but known for teaching talented children. He excelled and graduated with honors. His loves were math and chemistry, and he was not a bad athlete either, earning a position on the team's gymnastics program.

One of Carroll's notebooks from around that time (1930), was filled with hand-written notes and drawings of his experiments, most having nothing to do with gemstones. After his father's stern warning about diamond research, young Carroll turned his attention to other precious gemstones. He learned that French chemist Auguste Verneuil developed flame fusion methods to create ruby in 1902 and sapphire in 1909.

There was also apparently a lot written about the attempts to grow emerald, the next important gemstone. He read about the work of the I.G. Farben in Germany and its product called Igmerald, a duplicate of natural emerald that was not commercially viable. Farben did not explain how it grew its emeralds, so Carroll continued to dig, spending numerous hours in the Mechanics Library in downtown San Francisco. He could not find any examples of someone who had created a commercially successful emerald material.

What he thought was an easy few weeks project turned into a lifetime quest.

A glass vial of some of the first emeralds made by Hoytfield & Perry in France in 1847. Although they had no commercial value, they were a laboratory first in growing emerald crystals.

Carroll's first lab in the cellar of his father's house when he was around 12 years old.

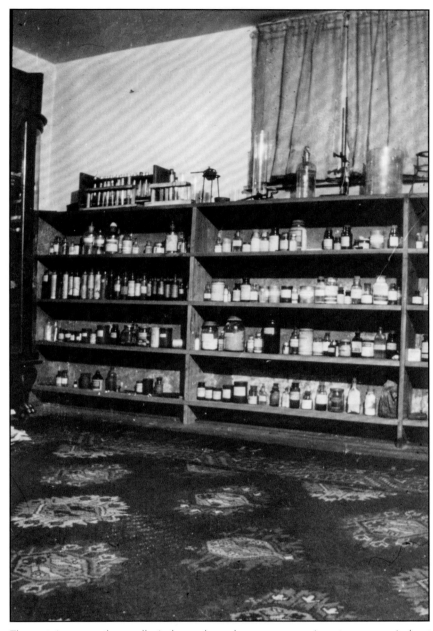

The curtains cover the small windows where dangerous experiments were carried out.

3
The First Chatham Laboratory-Grown Emerald

Carroll came home on his next break to find his furnace cold ... having nothing to lose since the cooling had caused the crucible to bulge and split, he hit it with a hammer. To his surprise, there were green crystals on the walls and bottom of the crucible. Carroll had made his first emeralds!

Carroll, 1947, in his first lab using glass apparatus he made by hand because what he needed was not available in local shops.

Carroll graduated from high school in the summer of 1931 and was accepted at The California Institute of Technology (Caltech), in Pasadena, some 500 miles from his hometown of San Francisco. Before he left for his semester at school, however, he began his emerald research.

Emerald's composition was schoolbook knowledge: beryllium, aluminum, silica and a little chromium for color. Just melting these chemical elements together to create emerald doesn't work because they have incongruent melting points: as one begins to melt, the other boils off. So, Carroll needed to find a flux material they would all dissolve in, but that would not become part of the resulting crystal. The same concept applies to sugar in hot water: the sugar "dissolves" in the water but does not "melt" and when controlled correctly, you can grow sugar crystals.

The first furnace Carroll created was made from a discarded 20-gallon strawberry bucket that he lined with refractory clay and a simple wire resistance heater he designed for the bottom. His first crucible, about the

 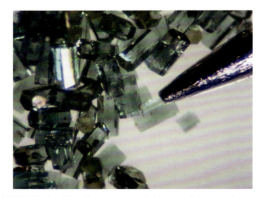

Left: Some of the first beryl crystals Carroll grew in 1933. **Right:** The first successful emerald crystals, which grew by accident after my grandfather William turned off the furnace while my father was away at the California Institute of Technology in 1934. The metal bar to the right is a needle that shows the size of the crystals, which were too small to be faceted.

size of a coffee cup, was made of platinum because he knew the chemicals he was going to melt would easily eat through any other metal at a high temperature. Platinum was relatively cheap in the 1930s because no real uses for it were developed yet. How and why he chose the chemicals he did is both a mystery and a closely guarded secret to this day. With that said, he still couldn't get his molten brew to make anything except a mess.

One fateful event changed his work forever.

He set up his umpteenth experiment during a semester break from college. He knew the exotic and corrosive nature of the solvents he was cooking at 1,000 Celsius would eat through many materials, including his heating element, so he wired a light bulb into the circuitry of the power source of the element. The light bulb would indicate to him that all was well in the heating department. If the light was on, all was well. If the light was off, the element had burned out.

He turned the power on and returned to Caltech.

Carroll's father knew nothing of the things going on in his basement; he was into lumber, not chemistry. One day he saw a light on in his son's lab room and turned off the power in a sense of conservation: "Damn kid left the light on." When he turned off the light switch, he turned off the power to the experiment.

Carroll came home on his next break to find his furnace cold. Figuring that the element must have burned out, he took the little strawberry bucket furnace apart, but found only undamaged parts. Next, he looked at the crucible with rock solid flux material inside and having nothing to lose since the cooling had caused the crucible to bulge and split, he hit it with a hammer. To his surprise, there were green crystals on the walls and bottom of the crucible. Carroll had made his first emeralds!

I am sure he was excited but perplexed at the same time. Why did the emeralds form? What was different in this run? He tried it again and again and got nothing! Why did the emeralds grow that one time? These same questions and topics have plagued our labs for 80 years: something so simple, so stupid even, that you don't consider its importance.

It took Carroll three years to figure out

One of the first emerald crystals Carroll grew. This one is quite attractive standing about 1.5 inches high. Credit: Tino Hammid.

Carroll Chatham's first business card.

that his father's act of turning off the power allowed the solution to cool slowly and from this supersaturated solution of emerald constituents, beryl crystals precipitated out of the solution and grew on the walls and bottom of the crucible.

This was around 1936, when Carroll was 22 years old. Once that small but significant action was determined, he never had another complete failure. He was able to make lots of low-end crystals, but they were all emerald. He theorized, correctly, that a temperature gradient was required to prompt the crystal growth. Another bit of help was starting with low grade natural beryl as seeds.

Carroll's accomplishment began to circulate among scientific circles. Front page articles ran all over the world and even *Ripley's Believe It or Not* ran a short blurb on his amazing accomplishment.

Classmates at Caltech, such as Linus Pauling, implored him to write a paper on his work "for the good of society" and Caltech tried to claim ownership because of his education at their school. Carroll said no to both – he was not into scientific fame, and he had proof of his prior work at home before entering Caltech. This was not the last time a company tried to take away ownership of his discovery.

My father may have grown up in privileged surroundings, but he was anything but spoiled. His father was a tough old man and sorry was the day any of the help talked back to him. Carroll had no interest in the lumber business and did not expect any benefits from it to come his way.

He saw his emerald crystals as a future form of employment – a job to earn a living.

4
The Birth of Chatham Research Laboratories

Carroll took one of the larger crystals to one of the few lapidaries in the United States, right in San Francisco on Geary Street. Francis Sperisen was amazed when my father told him he had grown the crystals he placed on the counter in front of him ... [and] was surprised when he began to work on one: "They cut just like real emerald!"

Carroll in front of his first professional lab in the outer Mission district of San Francisco, which was a good fit for a lab – not too industrial and not too residential. Millions of carats of rough emerald were produced here (in an area of about 4,000 square feet) until it closed in 2015.

After graduating from Caltech in 1937, Carroll got a job as a food chemist with the California Packing Corporation (Calpak), which was later to be bought out by the Del Monte Corporation. His job was trying to perfect ways of preserving foods in cans that didn't ruin the contents or kill the consumer. This sounds easier than one would think. Tin cans react with a lot of foods and the United States was moving into the mass marketing of foods. The only way to do that at this point was to freeze them, as in meat, or can them as in grandma's jellies and jams. Neither was a practical solution for mass marketing.

Carroll hated the job and his boss, but had no other choice at the time, so he made the best of it – and ended up making some advances in the preservation of canned food. I can't give him full credit for this accomplishment because he was part of a much larger team of chemists working on this and other problems connected to the food industry. Organic chemistry was not his specialty, but I'm sure he contributed something toward the success of the canning business.

He also continued his emerald research in his father's cellar. The crystals were getting bigger and better. The crucibles had grown, and he created temperature controls to keep things steady. He purchased crude electrical temperature readers at second-hand stores and worked to adjust the temperature gradient between the hot part of the melt to the slightly cooler

Original temperature controllers, 1936.

This temperature measuring device could read up to 1,500 Celsius using a platinum-iridium thermocouple. It was used until 2015.

growth section. This is something we still experiment with today.

His bosses at California Packing heard of his scientific exploits and tried to claim ownership, but again he was able to prove prior knowledge, so they were shut out – and not too happy.

Carroll took one of the larger crystals to one of the few lapidaries in the United States, right in San Francisco on Geary Street. Francis Sperisen was amazed when my father told him he had grown the crystals he placed on the counter in front of him.

"Can I facet one?" Sperisen asked. "Please do," was my father's retort, because that is why he came there in the first place.

Sperisen was surprised when he began to work on one: "They cut just like real emerald!" was his first reaction. The two men became close friends and business partners. Sperisen would do the cutting and sell from his shop; my father would also try to sell some on his own.

Carroll was 26 years old in 1940 and had avoided the draft due to college, but soon after graduating, he got Uncle Sam's "We Want You" message. Since he was working in a sensitive area as a food chemist at the time, he was able to continue his work at the California Packing Corporation as his military duty.

He was happy to be researching peaches and other fruits, when war broke out a few years later after the attack on Pearl Harbor, even if it pinned him down to his job as a food chemist. His work became extremely important for the war effort; Troops may need ammunition, but they also must have food and canned peaches quickly became an infantry man's favorite.

One story he shared with me about those canned peaches happened around Christmas time one year. He and the other scientists decided to give the troops a little present, so in addition to the sugar and water in the canned peaches they spiked a few hundred cases with pure grain alcohol. I'm sure they made some G.I.s happy that Christmas.

Carroll met his wife, Barbara, through a mutual acquaintance her older sister was dating at the time. They married in 1940 and rented their first home, as opposed to buying because my father's job paid so little and borrowing from Grandpa Chatham was out of the question. The home was far from palatial on a $200 a month salary – just another row house about five blocks from his father's grand brick house. My brother John was born in 1943, then I was born in 1945. They bought their first house in 1950 for $8,000 cash.

One of the first furnaces built around 1938 used a Wheelco temperature controller.

These platinum baskets hold seeds and flux, post growth.

By 1944, Carroll decided to build a laboratory in the outer Mission district of San Francisco, at 14th St. and Folsom Street, an area of light industrial businesses mixed with personal residences. He borrowed $10,000 from his father and built a 2000-square foot wood-frame building with no windows, just a front and back door. (It was more acceptable to borrow money from his father to build a lab as a business investment than a house to live in, which was considered a luxury by the standards of the day.) This was strictly working space; no office to sit in and the bare walls covered in fireproof Masonite. All the work benches were the same with the worksurfaces made of fire-resistant asbestos board, a great material for handling – or mishandling – caustic and/or hot materials.

The first lab had 10 furnaces to start, then doubled to 20 as the business grew. Carroll placed a fume hood in one corner for boiling acids and 20 furnace cans the size of garbage cans were ordered. He installed oversized electrical panels made to handle 400-volt inputs and a heating muffle that was built to go to 1,500 Celsius. Wheelco electric temperature controllers were ordered, two per furnace, for over-ride protection. Massive transformers were mounted behind every furnace to regulate the incoming current and to stop power swings, which were common at the time. Typically, anywhere from 109 to 130 volts came in, but if the huge machine shop down the street powered up everything, voltage would drop to 60 and the lights would momentarily dim. All these variables had to be considered when growing crystals at 1,000 Celsius, a one- or two-degree fluctuation would affect the quality. Every time we had a dip in power, a fine line could be seen in the finished crystals.

Three great things happened in 1945: the war was over, I was born, and Carroll quit his day job at the packing company. He also paid his father back the $10,000 he had borrowed.

```
                              November 1, 1947

              CHATHAM EMERALDS

                         Mfr's Price   Whlse.Price     Retail
1st Quality, Em. Cut . . . . . @ $ 30.00 per c  $ 45.00 per c  $90.00 per
(Clear)                                                               c
2nd Quality, Em. Cut . . . . . @   20.00        30.00          60.00
(Flaws or cracks, partly
  muddy - brilliant)
3rd Quality, Em. Cut . . . . . @   10.00        15.00          30.00
(Slight brilliance)
3rd Quality, Cab. Cut . . . . . @   8.00        12.00          24.00
(Slight brilliance)
Student Quality, Em. Cut . . . @   5.00          7.50          15.00
(Med. Green, no brilliance)
Student Quality, Cab. Cut . . . @   4.00         6.00          12.00
(Med. Green, no brilliance)
Specimen crystals . . . . . . . @   1.00         1.50           3.00
(Completely muddy)

Matched Calibre Stones - Clear

1.0 mm and less                  $ .50 ea.     $ .75 ea.      $1.50 ea.
Increasing $.08 per .1 mm to 2.5 mm
2.5 mm                                          1.70 ea.       2.55 ea.     5.10 ea.
Matched pairs - Clear, on order   Cutting Charge  Mfr.x1½ ea.  Whlsex2"
Less than 2.5 mm                  Plus $.75 each
2.5 mm and over                    40.00 per c   60.00 per c  120.00
                                                                  per c
```

Carroll's first price list, 1947.

The end of World War II released Carroll from his obligation to work at the canning plant and he couldn't get out quick enough. He was probably a lousy employee anyway, because he was always thinking about his other job: how to grow emeralds. He told his boss, "I am here to announce my resignation from this company from this day forward. Unfortunately, the meager salary I am paid will not cover my income taxes I owe from other earnings!" His boss was upset but could do little about it.

Carroll worked full time constructing his laboratory on 14th St. and by 1946-47, it was in full production growing emeralds. Many newspaper articles were written about his endeavors and magazine writers were constantly beating on his door.

A 1947 *Fortune* magazine article on my father was the inspiration for another crystal grower,

No admittance to Carroll's lab was strictly enforced - even for me - until I was trained.

Pierre Gilson. According to his son, Pierre Jr., the article inspired Gilson's search for growing his own emeralds.

Many people complimented Carroll on his accomplishments, but some condemned him. One person suggested that he was a German spy who had stolen notes from the I.G. Farben in Germany, where they were trying to grow emerald during World War II. Another accusation came out of Europe, calling Carroll Chatham a fraud. The FBI came around and asked questions, since we had just beat Germany in the war and people were very sensitive to anything connected to Nazism. It was easy to verify my father's years of research, his history of accomplishments – not to mention the fact that he had never left the United States in his life. That was easy to prove.

Not so easy was overcoming the threat many perceived Carroll's discovery could mean to their jewelry and gemstone livelihoods.

Carroll Chatham examines a pile of freshly harvested emeralds for an article in *Science Illustrated* written by Frederick H. Pough in 1948. Credit: Jon Brenneis and David B. Eisendrath, Jr.

Carroll Chatham: "The first and only successful grower of emeralds in the world!"
Credit: Elise Hix / *Ripley's Believe It or Not!* magazine, 1951

CHATHAM: Spends his mornings answering letters, balancing books.

Quest of Kings

When the California Federation of Mineralogical Societies opens the 16th annual convention of the International Gem & Mineral Exposition in San Francisco next month (July 8-10), all eyes, once again, will be on an emerald, especially grown for the meeting by Carroll Chatham of Chatham Laboratories. Reason: after a series of unsuccessful starts and stops, Chatham has his synthetic emerald business going at a 5,000-carat/month clip. (In dollars, production brings in about $25,000/month on a jeweler's retail basis.)

Further: the emerald Chatham plans to show at the Gem & Mineral Exposition will reportedly be the third largest ever produced synthetically—something under 1,000 carats.

Fitful Background: Background of the rags-to-riches story goes back to Chatham's boyhood and a hobby interest in gemmology. His first stone (bought by a San Francisco dealer) was sold in 1940. But the path to success hasn't been easy. Jewelers in the U. S. weren't easy to sell on the practicability (or salability) of Chatham's synthetic emeralds; on a number of occasions, Chatham has been investigated by both the Federal Trade Commission and the Jewelers' Vigilance Committee.

Making matters even more difficult: Chatham has constantly refused to seek patent protection on his synthetic emerald production process. He (to the present date) has never incorporated, and his wife is the only other person ever admitted to that part of Chatham Laboratories where the autoclaves and other processing equipment are located.

Nevertheless, production today is on a fairly steady keel; and markets are opening up in all sections of the world, including India, Japan, most of Europe, and North Africa. Crystals, SA, of Switzerland, has first option on 90% of Chatham Laboratories' emerald production, and has continually exercised its option since the agreement was finalized in 1951.

One-Man Job: From the standpoint of actual time spent on the production line Chatham's day reads like a movie scenario. On a typical weekday, he rises late, spends the morning working on correspondence, finances, etc. Then after a leisurely lunch, he goes to the laboratory to read more mail, check on his process, clean up rough stones and ready them for shipment. The synthetic emerald process itself, Chatham claims, requires only one hour of his time per day on the average. And if he chose to spend the money to install automatic equipment,

PAPER WORK OVER: Visits a San Francisco lapidary, gets to the office sometime before noon.

Chemical Week magazine's "Quest of Kings" article portrayed Carroll Chatham and his wife Barbara's typical work day in their research laboratories in 1955.

5

Guerrilla Marketing, Chatham-Style

Little by little word got out about my father's emeralds. Major magazines called for interviews, and he received an enormous amount of free publicity. Periodicals such as *Reader's Digest*, *Fortune* and *True*, numerous newspaper stories and even *Ripley's Believe It or Not!* featured articles on him. Slowly, very slowly, this ground-roots style of guerilla marketing started to pay off.

Growing emeralds is quite an achievement for any individual. My father, however, soon found out that the masses do not beat a path to your door for inventing a better mouse trap – you must sell it!

Marketing was not one of Carroll's strengths, not even close. So, with hat in hand and his wife at his side, he took a few hundred cut emeralds to New York, the world's epicenter for jewelry and gemstones during that era. If you were anybody in the jewelry business, you had to have an office in Manhattan, preferably one close to 47th St. and Fifth Avenue, the hub of all gemstone cutting, colored stone trading and jewelry manufacturing.

My father was an honest man. He could have quietly sold his stones as natural, and no one would have known the difference. It would have been identified as emerald by any laboratory, but it would have certain characteristics that set it apart from Colombian or Brazilian emeralds; it would be identified as one from a "new" mine of an unknown origin. He could have reaped millions and there were plenty of people who wanted to help him do this "correctly," including some mine owners.

But Carroll wanted his stones sold for what they were: the Chatham Cultured Emerald. He even considered protecting his process with a patent and went to an attorney specializing in such law. After listening to my father's story and looking at the gemstones, the attorney wisely concluded it would be a waste of money. "Why patent the process when no one can tell by looking at the stones how they are grown? Keep it a secret, that is the best patent there is."

And so, it has been a family secret to this day. No one has figured out how Carroll Chatham could grow emeralds.

As my parents walked down Fifth Avenue, they saw jewelry worth millions shown in the plate glass windows of the finest and most exclusive jewelry stores, including Harry Winston, Tiffany & Co. and Cartier. They went into Van Cleef & Arpels and asked to talk to the manager to see if the store purchased loose gemstones.

"Perhaps," was the typical evasive New York answer, "what do you have to sell?"

Carroll took out several envelopes (called diamond papers) of the type used in the gem trade and proceeded to display hundreds of cut stones of the very best quality found in natural emerald. These stones ranged from 1/2 carat to 2 carats in size. The manager looped the stones, one after the other, confirming to himself they were indeed emerald.

"Where did you get such fine emeralds?" the manager asked.

"I grew them in my laboratory in San Francisco," Carroll proudly stated.

"I see," said the astonished manager. He then said to his assistant, "*Appelle la police.*"

My mother, a graduate from U.C. Berkeley, spoke fluent French and told my father, "Carroll, he just told them to call the police, pack up and let's get out."

This was Carroll's first marketing blunder of his budding career: He misjudged the perception of the jewelry trade's attitude of anything laboratory grown and started at the top of jewelry stores, such as Van Cleef & Arpels, when he should have started at the bottom and worked his way up. They never went to one of those types of stores again. My father was finding out it was hard to tell the truth and make any money.

This marked the development stage of his marketing education, of which he knew little. He reached out to the rockhound community, such as the Gem and Mineral Society in San Francisco, where he would display his crystals and cut stones and give an occasional talk. He also reached out with small, simple advertisements placed in various lapidary magazines that were very popular in those days. In addition, his friend Francis Sperisen,

the gemstone cutter, advertised Carroll's emeralds in magazines devoted to hobbyist and gem collectors.

Little by little word got out about my father's emeralds. Major magazines called for interviews and he received an enormous amount of free publicity. Periodicals such as *Reader's Digest*, *Fortune* and *True*, numerous newspaper stories and even *Ripley's Believe It Or Not!* featured articles on him. Slowly, very slowly, this ground-roots style of guerrilla marketing started to pay off. Even with all that publicity, the world of Van Cleef and Tiffany was not our target market and never would be.

I have thick files of personal correspondences from hundreds of individuals and companies wanting to know more about this "guy who could grow emeralds." Many asked for donations, many wanted to know where they could buy his stones and some very large companies, such as Union Carbide, asked for meetings. Luckily, my mother was an excellent typist so everything was carbon copied and filed. It is interesting to read through these requests and unless the writer was a real crackpot, so indicated on the letter as 'not ans' [not answered], Carroll would write them back and answer their questions as best he could.

Just to give you an idea how prehistoric the gemological world was back then, I have personal correspondences from 1947 from Richard T. Liddicoat, then-director of education of the Gemological Institute of America (GIA), which was training jewelers to become professional gemologists, asking Carroll for any suggestions he may have on separating his emeralds from natural. There was no way the average jeweler, without a microscope or training, could possibly identify the stones. Gemology was in its infancy, and no one understood what a gemstone actually was.

Carroll featured in *Daily Palo Alto Times*, 1947.

There were no requirements or credentials to becoming a jeweler back in 1947 and unfortunately there are still none today! Many organizations offer training in gemstone identification and jewelry making, but I have seen many graduates of these programs, including some of my employees with diplomas, forget everything once out the door. There should be yearly retesting such as the International Society of Gemologists, run by Robert James in Texas, does. If you want to learn gemology that will stick for life, go to www.schoolofgemology.com and take their courses. You won't regret it.

Correspondence between Richard T. Liddicoat, director of education at the Gemological Institute of America, and Carroll Chatham, concerning the ability to distinguish Chatham's laboratory-grown emeralds from natural emeralds.

Richard T. Liddicoat at his desk in the spring of 1946. Credit: GIA.

"Since your product is so beautifully made and compares so favorably to fine natural material, we doubt that it would be possible to teach a method of identification by recognition of the differences between inclusion in your product and in natural emerald."

Richard T. Liddicoat,
Director of Education
Gemological Institute of America

Gemological Institute of America
(UNITED STATES AND CANADA)
A Non-profit Educational Institution
International Headquarters
541 SOUTH ALEXANDRIA AVENUE
LOS ANGELES 5, CALIF.

Date December 11, 1947
VIA AIR MAIL

Mr. Carroll F. Chatham
70 Fourteenth Street
San Francisco 3, Calif.

Dear Mr. Chatham:

I read with interest the article on synthetic emeralds in "Jewelers Circular Keystone." I was particularly interested in the paragraphs which Frederick Pough had quoted from your communications to him. I was quite surprised by your relegation of the fluorescence test "to the garbage can." I am sorry that neither Dr. Bandy nor myself was at the Institute during your visit of last summer so that we could have discussed this with you. Since that time we have checked numerous natural emeralds from every important world source and have yet to see any but very weak light red fluorescence and this only in two natural emeralds.

Admittedly, there is a possibility that a new emerald source in the future will produce emeralds which fluoresce as strongly as your present product. I have no doubt that if you wish to do so, you could make emeralds that fail to fluoresce. However, at the present time, synthetic emeralds do fluoresce and we have yet to see or hear of a report of natural emerald that has a sufficiently strong fluorescence to cause confusion to a user of this test.

Since your product is so beautifully made and compares so favorably to fine natural material, we doubt that it would be possible to teach a method of identification by recognition of the differences between inclusions in your product and in natural emerald. While the natural gem sometimes shows three-phase inclusions and crystallized pyrite which are not present in your synthetics, failure to find such inclusions is certainly not proof of a synthetic. Most jewelers do not possess magnifying instruments of high enough power to resolve the tiny three-phase inclusions. A test based on fluorescence is not one that we like to use if it can be avoided. However, until such time as fluorescent natural emeralds are found or synthetic emeralds appear which fail to fluoresce, it seems like the best test available to the average jeweler. If you have seen natural emeralds that fluoresced strongly under 3500 Å or have produced synthetic emerald of a quality comparable to your usual material which failed to fluoresce, we would like to know about it. Naturally, we would be ready to call this to the attention of our students and readers of "Gems and Gemology" immediately.

With best wishes for a happy holiday season.

Sincerely yours,
GEMOLOGICAL INSTITUTE OF AMERICA
R. T. Liddicoat
Richard T. Liddicoat
Director of Education

Would appreciate your reaction to the above.

Liddicoat to Chatham, Dec. 11, 1947.

> December 17, 1947
>
> Dr. Richard T. Liddicoat
> Gemological Institute of America
> 541 South Alexandria Avenue
> Los Angeles, 5, California
>
> Dear Dr. Liddicoat:
>
> With regard to your letter of December 11, I am sorry the statements presented in Dr. Pough's article were misinterpreted. My statement concerning the fluorescence test on the synthetic emerald was merely meant to emphasize the fact that the test was not infallible. Speaking as one scientist to another, the important thing in any determination, you will agree, is the cause which produces the effect. Any red fluorescence shown in the natural or synthetic emerald is due to the chromium content whether it shows strong or weak. Therefore, since both contain the same element in varying degrees, one must be very careful in proposing a confirming test without making certain that an exception is highly improbable.
>
> With the ruby or sapphire it is entirely different. Here one has a method of crystallization far removed from that produced by Mother Nature to such an extent that the boule can hardly be recognized visually as a single crystal and the well known characteristics show up and are easily explained because of the typical behavior of the torch.
>
> By all means, however, the fluorescence test is the most practical yet presented and gives an excellent indication of the origin of the emerald, especially when one has a quantity to classify.
>
> Hoping the above explanation will be of some assistance, I remain
>
> Yours very truly,
>
> Carroll F. Chatham
>
> CFC/bg

Chatham to Liddicoat, Dec. 17, 1947.

Carroll F. Chatham examines one of his laboratory-grown emeralds Credit: Fred Lyon.

"By all means, however, the fluorescence test is the most practical yet presented and gives an excellent indication of the origin of the emerald, especially when one has a quantity to classify."

Carroll F. Chatham, Chatham Emeralds

The subject of my father's emeralds made national television as well. A program called "You Asked for It," hosted by Jack Smith, had my father on for a live interview in 1955. This program was very popular, on a par with "60 Minutes," and someone from Topeka, Kansas wrote and asked: "Is there really a man in San Francisco who can grow real emeralds crystals?" I am not sure if these letter requests were real, but the topics made for good TV.

Jack Smith and a camera crew came to San Francisco, but since Carroll did not allow anyone in the production labs, he suggested a demonstration for filming in the workshop. "I am often asked how to separate one of my emeralds from a natural emerald," he explained. "I can show this demonstration for your audience, if you buy a natural emerald for the experiment."

Jack Smith sent someone out for a one carat natural emerald of good quality from the trade.

Smith interviewed my father, who was a nervous wreck; he never got used to public speaking and hated it. Carroll introduced his demonstration by saying, "Jack, I am often asked, how can you tell the difference between a Chatham stone and a natural stone? I have two emeralds here being held by a thin wire. I am going to prove beyond a doubt which one is a Chatham, and which one is the mined stone."

Two cameras were filming, (remember, this is live TV) with one camera on the two stones and one on Jack Smith. My father lit a torch and began to heat the two stones as Jack watched. Suddenly, "BANG" and the natural emerald exploded. Jack Smith gasps, seeing the emerald they had just bought go up in smoke.

"Jack, I am often asked, how can you tell the difference between a Chatham stone and a natural stone? I have two emeralds here being held by a thin wire. I am going to prove beyond a doubt which one is a Chatham, and which one is the mined stone."

– Tom Chatham

"Jack, I just proved that the stone that exploded was the natural stone because it has water trapped in the structure," Carroll said. "The Chatham stone is devoid of any water and is grown at 1,000 degrees Celsius, so the torch won't hurt it. Any family jewelry using Chatham Emeralds are safe in case of a fire burning down your house!"

Jack Smith, still staring at his pieces of stone on the bench, says "Ah, well, that's great Mr. Chatham. Thank you – I think."

These and other TV shows, magazine articles and newspaper stories were chiefly responsible for getting the word out about lab-grown emeralds. They let the consumer know there was a choice between a natural emerald and a laboratory-grown emerald – and that they were identical in every way. We would not be where we are today without this priceless exposure, all free of charge!

The television show **"You Asked for It"** taped a live segment of Carroll Chatham in his workshop after a viewer from Topeka, Kansas asked: **"Is there really a man in San Francisco who can grow real emeralds crystals?"** The host Jack Smith and his crew filmed a demonstration of Carroll heating two stones (a natural emerald and one of Chatham's emeralds) with a torch – until the natural emerald exploded!

6

The Demise of the Loop Lumber & Mill Company

My father was asked numerous times for infusions of cash for the [Loop Lumber & Mill Company] business and whenever his older brother called up to ask if he wanted some ducks he had just shot, he knew what was coming. "Uncle Bill is coming over for dinner, a $20,000 duck dinner!"

A Chatham family photo taken in 1947. My father Carroll is front and center with his father and brothers behind him. From left: my uncle Russell, my grandfather William and my uncle William Jr.

Carroll's two older brothers, William Jr. and Russell, had taken to the family's lumber mill business early in their work life, but struggled with it while Carroll was trying to market his emeralds. Although demand for wood was high in the war years, certain qualities of hardwoods were out of their reach. The Loop Lumber & Mill Company mainly cut clear heart redwood that was used in constructing houses, a need that dropped to nothing during the war.

The military needed stronger woods for rifle stocks, tent poles, wheel spokes, broom handles, which were not found on the West Coast. Times were tough in the lumber business with many mills sitting unused and shut down. It is estimated that out of 15,000 mills across the United States, only 5,000 were in production during the war.

Labor was also becoming more expensive, thanks to Harry Bridges and his new union, and transportation of materials was also a bottleneck. Cutting redwoods near the ocean and guiding them down the coast with a ship was easy and inexpensive. Cutting inland, as forests were depleted, meant roads had to be built, heavy machinery purchased and labor to run

A modern map of San Francisco showing the Chase Center (marked with the red circle). Travel back 50 years ago it was the former Loop Lumber & Mill Company yard. Credit: Adobe Stock.

them hired. It became harder and harder to be in the lumber business on the West Coast and in many other places in the United States as well. The lumber business was floundering for the Chatham brothers.

William had no doubt retired from the business by this time, so maybe he did not see the slow decline of revenues that his once-thriving business represented and did not witness the growing problems. By the mid 1950s, yard bosses were paid $50,000 a year and the price per foot of lumber was falling below the cost to cut it. Loop Lumber turned to other exotic woods, but it was too little too late.

I clearly remember my last trip to Grandpa Chatham's house in 1950. We all stood around his bed because the end must have been near. I had no idea what was going on because I was only 6, but I knew to be quiet.

Grandpa Chatham reached out and took my hand and palmed me a quarter, a huge sum at the time. He died days later at 88 years of age.

The family soon discovered that William died without a will, but the business was incorporated and had stock certificates, all in William's name; the house was free and clear. Back then, people didn't usually get loans from a bank, they used cash to buy things and because my grandfather strongly believed in this principal, he was cash poor. He built his house for cash, bought his car for cash and ran his business on cash. Everyone in my father's family all learned from the knee of William: "If you can't afford to pay cash, you can't afford to buy it!" How soon we all lost that common sense of thrift.

The house sold for $25,000, not even twice the price it cost to build it 30 years prior and

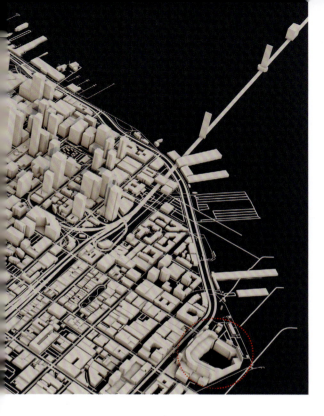

none of the Chatham children could afford to maintain it alone. The house sold and the lumber business continued to be run by the older brothers, and a deal was worked out between the four siblings. When the courts settled his estate, the stock was split four ways.

Soon a big problem loomed over their heads: The Internal Revenue Service (IRS) came around for their pound of flesh. The IRS collects inheritance taxes on the total value, regardless of the form it is in. The stock of the lumber company reflected a business with a rich past but a poor future and the IRS used a formula based on past revenues to come up with a value of $400 per share of the business, of which a certain percentage had to be paid in inheritance tax.

My father became the lead person in the fight with the IRS and told them, "If you think this is worth $400 per share, I will sell it to you for $200 per share." The IRS, of course, doesn't work that way.

Taxes in the 1950s were very high, about 92%, plus state taxes over a certain level, and the number the IRS came up with was way over that. The company was losing money and barely keeping its doors open, but the taxes were paid, leaving them all the poorer.

An important lesson was learned by all. Never die without a will or better yet, a living trust to protect those you leave behind. It was crippling to William's four offspring.

My father was asked numerous times for infusions of cash for the business and whenever his older brother called up to ask if he wanted some ducks he had just shot, he knew what was coming. "Uncle Bill is coming over for dinner, a $20,000 duck dinner!" he would announce to us all. I'm sure Aunt Ruth got the same requests, and both tried to be as helpful as possible, but they knew this cash drain had to stop. They had no idea what was happening in the lumber business but wanted out.

In the late 1950s they made a deal with the older brothers: sign over the ownership of the land and keep the business and any profit it may make. And so it came to pass, the mill land and docks in Alameda and the storage land in San Francisco was deeded over to my father and aunt.

The business struggled through the 1950s and finally closed its doors in the 1960s. The land sat idle for many years, located in a very undesirable section of the San Francisco waterfront. Eventually, Carroll and his sister sold it to a large land developer, which also let it sit idle for many years. The area was finally developed 50 years later, and our former lumber yard is the home of the San Francisco Warriors basketball team – Chase Center (if only we held on to it!)

7
A Swiss Partner Swoops In

Being connected in the New York jewelry industry was immensely helpful to get things started. You could call up one of your buyers and say, "Hey Bernie, I have this brand-new product called Chatham Cultured Emerald. We have an exclusive on it and we want to give you the first shot at manufacturing a finished product for us. We will advertise it in national magazines, and we want you to put it in your current line in stores."

Dan Mayers, a geologist from Europe, came into my father's life sometime around 1953-55. He had heard about my father's emeralds all the way over in Switzerland and asked for a meeting to learn about them. He flew over and they met in our home because my father did not have an office outside of the lab to receive guests. This meant mom and us kids often sat in on some of the meetings. We could listen but had to keep our mouths shut.

Dan had a dark Indian complexion and was always smacking his lips, like he was about to eat something. He didn't drink or smoke and loved to talk about himself; he seemed to know everything about anything, saying he graduated from Harvard. My father was impressed with this man of the world and it became clear that they were talking about working together.

I did not hear all the negotiations and it took many letters and visits to cement the deal, but the outcome was this: Dan would take the entire production of Chatham emeralds on an exclusive basis, except for a few special stones he would allow Francis Sperisen in San Francisco to cut and sell. My father would grade the rough stones, based on an agreed grade sample and price them. A bill would be submitted, and funds transferred into Carroll's account before shipping, an unheard-of practice in the gem trade.

My father was ecstatic. Now, he could devote all his time to research and production and not worry about marketing.

Dan promised to sell the stones legitimately and developed a marketing plan in New York to do so. He found a firm called the Ipekdjian Bros., run by Adom and George Ipekdjian, diamond brokers right off 47th St. in New York City. They were in good standing with the trade and had an advertising agency, owned by Edward Coyne, working within their offices. The Ipekdjians took care of their diamond department and Coyne took care of the advertising of the Chatham emerald account.

Carroll's only stipulation was that he had final say on any advertisements – he wanted to make sure no one got a little too creative in writing about the emeralds – but since he had no ownership in Ed's or the Ipekdjians brothers' business, they felt little need to consult with him. Dan didn't seem to care one way or the other and much to my father's dissatisfaction, he did not see new ads or brochures until after they were printed or distributed.

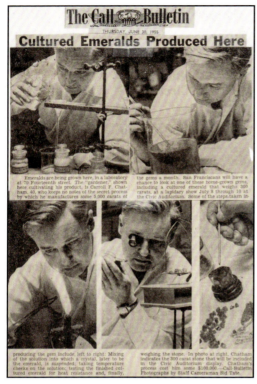

Carroll Chatham and his "cultured emeralds" were featured in the *The Call Bulletin* newspaper in San Francisco in 1955.

The Ipekdjian brothers were only interested in the diamond business, so they let Ed Coyne do what he wanted with the Chatham account. Dan did nothing or even saw the products we produced, since he lived in Switzerland. Not being privy to anything in that partnership does leave some unknowns about ownership and profit sharing. It was a strange business arrangement for sure.

Being connected in the New York jewelry industry was immensely helpful to get things started. You could call up one of your buyers and say, "Hey Bernie, I have this brand-new product called Chatham Cultured Emerald. We have an exclusive on it and we want to give you the first shot at manufacturing a finished product for us. We will advertise it in national magazines, and we want you to put it in your current line in stores." Who you knew in the business was very important back then and to some extent is even today.

"Who you knew in the business was very important back then and to some extent is even today."

– Tom Chatham

Things went swimmingly well for Carroll's emeralds all through the 1950s.

Ed was doing a great job marketing my father's emeralds and was so successful, he created a new company called Cultured Gemstones, Inc., that was still partly owned by the Ipekdjian brothers. He oversaw all the cutting and distribution of the Chatham emerald, which was now a full-time job for him.

Stones were always in demand: production was at about 5,000 carats of rough a month. The company was still cutting everything in New York and produced about 1,500 cut carats a month, keeping about 20 cutters busy. Ed set up all the cutting of the emerald rough into sizes the jewelry manufacturers requested.

The checks poured in, the quality of the emeralds was improving and I remember my father contemplating another 4,000 square feet of laboratory space for additional furnaces.

Then disaster struck.

8

The Backlash of the Industry, Courtesy of the FTC

"I call them cultured because I start with a natural emerald seed and over the course of a year the crystals grow from natural materials containing the elements of emerald," he said. "The end result is emerald, not fake emerald. Anyone who hears 'synthetic' thinks fake, ersatz, imitation, etc. – and I refuse to call my emeralds fake or synthetic."

– *Carroll Chatham*

Some people took exception to my father's opinion that the Chatham emerald was superior to a natural emerald because the Chatham stone would not be damaged in a fire. The TV exposure must have been in the millions when "You Asked for It" aired in 1955. Plus, Ed Coyne's national advertising to the public about "the world's newest jewel" was finally paying off.

I am sure some political strings were pulled, but how we found out about the uproar was in a press release that the Jewelers Vigilance Committee (JVC) sent to all the jewelry publications on Jan. 9, 1959 (edited for repetitive legal language):

"At the request of the Jewelers Vigilance Committee … concerning the use of the terms "cultured," "created by man" and "man made," … in conjunction with precious stone designations, … the FTC [the Federal Trade Commission] expressed the following opinion in a letter recently received (by the JVC):

> … It is the opinion of this Bureau (FTC) that the words "cultured," "man made," and "created by man" are not words of like meaning with that of "synthetic" and that such terms should not be in conjunction with the words "ruby," "sapphire," "emerald," or the names of any other precious stones …

The press release went on to state that many firms and organizations indicated extreme concern relative to these matters and the importance of the JVC in protecting the buying public from statements and representations that may be misleading.

Obviously, Carroll did not concern himself with this development until a surprise visit by an FTC lawyer to his lab. He wrote about it in a May 4, 1959 letter to Mayers, who was in Switzerland.

"The FTC attorney appeared unannounced at the lab. I pointed to the sign directly facing him on the door: Hours by Appointment ONLY. He demanded entrance – I said, 'Come back Monday and I'll let you in.' "

Carroll called the FTC attorney at 8 a.m. Monday and switched his appointment to Howell Lovell's office, his long-standing family attorney. When the FTC attorney arrived, he pulled out a list of questions "a mile long" and said, "First and most important, I must know the details of how you produce the emeralds and secondly, I must inspect your equipment."

"And if I refuse?" Carroll asked.

"You don't refuse the Federal Government on ANYTHING, and if you try, you will be subpoenaed, along with all files, letters and records, and if you still hold back, there will be a fine and imprisonment."

Carroll laughed and responded: "Then comes the rubber hose, thumb screws, a brain washing and then what?"

At this point, Howell "jumped out of his chair and said 'Chatham, be quiet. I will do all the talking and interviewing from now on.' "

He proceeded to address the FTC attorney: "Listen fella, first of all, your intimidations do not scare either Mr. Chatham or myself. We will start at the beginning – and I don't mean by your first question, I mean the U.S. Constitution, Bill of Rights, etc., and what limitations the FTC has. In the first place, the FTC has no jurisdiction over anyone not engaged in interstate commerce. According to Mr. Chatham's contract with Anglomex, Inc., (owned by Dan Mayers) sales of all products are transacted, consummated, delivered, and even paid in advance at …. City and County of San Francisco, California. The goods are no longer Mr. Chatham's property the instant they leave his front door, even though he ships them to N.Y. … shipment is made by Mr. Chatham merely as a courtesy, nothing more."

My father described the FTC attorney as slumped in his chair "with a sheepish look on his face."

Howell continued: "So just put your long list of questions aside and we'll spend the next few years of this case on whether you have the right to pose any questions to Mr. Chatham in the first place, let alone threaten him, and secondly, whether you even have the right to be in my office."

The FTC attorney responded, "Pretty clever" and Howell proceeded to lecture him for four more hours.

The next week, the FTC attorney called Carroll asking for a private appointment because he didn't want to "waste Lovell's time."

Carroll's response: "Oh that's all right, Mr. Howell's got more time than I have, and he likes to talk to you."

Howell did meet with the FTC lawyer again and informed him that all important papers from Mr. Chatham were now in his vault and to learn anything more, the FTC would need to subpoena him and his files.

The FTC attorney's response: "Now you know we can't do that, it's against the law to violate attorney/client privilege."

"Well, your memory is getting better – or have you returned to law school?" Howell asked.

All of this back and forth took the attitude out of the FTC attorney, but Howell knew they had to do something, so they composed a long letter explaining Carroll's position and sent it to the head of the FTC in Washington, D.C.

"We tried to be most humble and cooperative to the main office in Washington, D.C., since there is no sense in locking horns unless we have to," my father wrote to Mayers. "After all, our main purpose is to attempt to make them drop the whole case."

Howell told Carroll that if they refused to cooperate "there are a dozen technicalities they can subpoena you on."

Many months transpired with little action, until Oct. 13, 1959, when the six-page Letter of Complaint, docket number of 7609 in the matter of Carroll Chatham, Dan Mayers (Anglomex), the Ipekdjian Bros. and Ed Coyne of

{41}

Cultured Gemstones, Inc. arrived.

Basically, the complaint outlined the advertising they found in violation of various statutes and gave Chatham and the others 30 days to answer the complaint or admit guilt by default.

"It has been brought to our attention that you are manufacturing a product and calling it 'cultured emerald'. Since there are no complaints from consumers, simply signing this order and amending your advertising will stop any further action. Please respond immediately."

Carroll and the others decided they had no choice but to fight the action.

FTC hearings take place in Washington, D.C. and only lawyers licensed there could represent Carroll Chatham. His San Francisco lawyers and New York lawyers represented Ed Coyne and the Ipekdjian brothers on the side. Dan Mayers and Anglomex were a foreign entity and not served, but he certainly had a financial interest in this legal proceeding. This was an expensive battle to fight in Washington, D.C., but Carroll, Mayers and Coyne and felt they had no choice.

A judge presides over the hearing and a panel of select FTC specialists are also present. Witnesses are called and questioned, under oath. Richard T. Liddicoat, Robert Crowningshield and Bert Krashes – all gemological experts from GIA – were called to give testimony, as was Reginald Miller, a noted natural emerald cutter in New York. Each was asked a series of questions about gemology and their respective experiences in that field. The questions eventually led to one question from the judge: "What does Carroll Chatham make, in your professional opinion?"

All answered, "Emerald, your honor," ac-

cording to the transcripts from the case. The judge turned to the government's council and asked, "Whose witnesses are these, the governments or Chatham's? We are not getting anywhere here."

My father was then called to testify and after he was sworn in, the opposition started to question him about his process and why he called his product "cultured emerald."

"I call them cultured because I start with a natural emerald seed and over the course of a year the crystals grow from natural materials containing the elements of emerald,' he said. "The end result is emerald, not fake emerald. Anyone who hears 'synthetic' thinks fake, ersatz, imitation, etc. – and I refuse to call my emeralds fake or synthetic."

The judge, who also held a degree in chemistry, then asked my father, "for the good of society and for us to make a determination of your claims, you must describe in detail your process and all the chemicals involved."

Carroll replied: "I refuse to divulge my life's work. The end results speak for themselves."

"Mr. Chatham," the judge said, "refusing to answer my question is in contempt of court. A fine of $5,000 per day can be levied on you plus a year in jail for every day you refuse to answer."

My father paused and then spoke.

"I refuse to answer, you can lock me up today. I will not tell you how I grow emeralds. I will not ruin my life's work for the good of anybody, especially not these people trying to put me out of business."

Silence. A long silence...

The judge had made a big legal mistake. You can't make someone give up a secret without

S. F. News-Call Bulletin
★★★★ Fri., Dec. 11, 1959

At Issue: Synthetic or Cultured?

S. F. Emerald 'Grower' Fights for His Secret

By ANDREW CURTIN
News-Call Bulletin Staff Writer

Carroll F. Chatham, 45, operator of San Francisco's unique one-man industry, has a secret and he's fighting to keep it.

The secret he's kept for nearly 25 years is how he "grows" emeralds in his Chatham Research Laboratory at 70 14th st. here.

CHATHAM IS defendant in a Federal Trade Commission complaint brought by a New York group of gem dealers, the Jewelers' Vigilance Committee.

Chatham is concerned on two points:

● Whether, in the course of FTC hearings scheduled here later this month, he will be compelled to disclose his secret process, which he has never patented.

● The possibility an FTC order might compel distributors of his gems to advertise them as "synthetic."

CHATHAM'S output, he said, is running to about 5000 carats a month, with a

CAROLL T. CHATHAM AND CULTIVATED GEMS
It required 10 months to "grow" this pile, valued at $50

The San Francisco *News-Call Bulletin* reported on Carroll Chatham's battle to keep how he grows emeralds secret in December of 1959. This was the beginning of the Federal Trade Commission due process of law…and no laws had been broken. Carroll had the right to refuse to answer.

They tried dancing all around this question, but every time they approached anything close to asking my father to explain how he "made" his emeralds, he refused to answer. He was on the stand for an exhausting eight hours a day for many days in a row as the government lawyers tried to get him to explain how he made his emeralds.

Ultimately, the judge assembled everyone in his chambers.

"I am not going to put you in jail, Mr. Chatham," the judge told my father, "but I want to end this circus, we are getting no place. Isn't there some other word or phrase you can use?"

"We have tried, your honor, we even hired a linguist named S.I. Hayakawa in San Francisco, and came up empty-handed," my father said. "Here is the list we rejected: re-crystalized, re-grown, re-constituted, manmade, etc. The list went on and on. Some were as problematic as cultured, and some were trademarked by others already."

The judge turned to the attorneys and asked, "What did we determine Chatham creates?"

All answered, "emerald."

"Why don't you call them Chatham Created Emerald?" the judge asked.

My father and his attorneys were shocked. "We never thought of that word, it's perfect. Will the court accept this?"

"Yes," the judge answered, "with certain provisos that legal will write up."

So, my father signed the "cease and desist" notice that ended his use of the term "cultured"

and finally, the official verdict was issued. On July 27, 1960, the FTC advised Carroll Chatham and the others that the commission had decided the term "Chatham Created Emerald" would be acceptable if not used unambiguously and that only the "emerald" had been created by Chatham and nothing else.

The FTC declared that from this date going forward, the product sold by Cultured Gemstones, Inc. (renamed Created Gemstones, Inc) in New York would be called Chatham Created Emerald, and that all advertising, corporate names, business cards, advertisements not yet in print would reflect that. All references to "cultured" must be dropped. It was also stipulated that no convolutions of the word "created" could be used, such as "Chatham Created Emerald Jewelry" that could confuse the consumer. If we failed to see that distinction, the action would be revisited. Case closed!

All respondents – Carroll Chatham, Ed Coyne, the Ipekdjian brothers and Dan Mayers – gave assurances that such care would be observed. All were overjoyed and relieved. It was over.

This sounds simple and straight forward, but it took six months and many trips back and forth to Washington, D.C., including a lot of legal expenses, to complete.

A month after the final settlement was written, all the corporate names had been changed. A lot of money was spent on one little word, but we complied. New brochures were printed, letterheads redesigned and printed, new ads were designed and advertising contracts were signed with the leading jewelry industry magazines. Many thousands of dollars were spent to change this one little word, from "cultured" to "created."

Everyone felt relieved and positive for the future. Chatham Created Emerald was born.

FEDERAL TRADE COMMISSION
WASHINGTON

RECEIVED
JUL 28 1960
AM 7 8 9 10 11 12 1 2 3 4 5 6 PM

OFFICE OF THE GENERAL COUNSEL

JUL 27 1960

Ipekdjian, Inc.,
and Cultured Gem Stones, Inc.,
580 Fifth Avenue,
New York, New York.

Attention: Adom Ipekdjian,
Georges Ipekdjian.

Re: Chatham Research Laboratories, et al,
Docket 7609.

Gentlemen:

I am authorized to inform you that on July 25, 1960, the Commission accepted my opinion that the respondents' use of the term "Chatham created emerald" does not violate the order herein unless used ambiguously, but that your compliance reports dated May 26, 1960 and May 23, 1960 with exhibits thereto attached was rejected by the Commission because of the manner in which the term "Chatham created" is used in certain places in the advertising, but when you report and submit advertising which makes it clear that it is only the "emerald" which has been created by Chatham, it will be considered in compliance.

As illustrative, you are advised that "necklace composed of Chatham created emeralds" would not violate, but "Chatham created emerald necklace" would. Great care should be taken to see to it that the words "Chatham created" are adjectives to and modify the word "emeralds" and nothing else. "Chatham created emerald jewelry" would be considered to violate this order. "Chatham created emerald ring" the same. "Ring set with Chatham created emeralds" would not. If it should develop that your copy writers are unable to perceive this distinction at all times, I shall suggest to the Commission that it give reconsideration to the matter.

Meanwhile, a further and satisfactory report showing compliance with this order along the lines herein indicated should be received within 60 days.

Very truly yours,

P. B. Morehouse,
Assistant General Counsel
for Compliance.

cc: Peter W. Quinn, Esq.,
270 Madison Avenue,
New York 16, New York.

A July 1960 letter from the FTC states that "Chatham created emerald" is an acceptable description for the Chatham product.

9

Whoops ... the FTC Changed its Mind!

The FTC doesn't tell you what word to use, it only judges on what you do use. The court got bogged down in semantics rather than what the JVC and others really wanted, "synthetic emerald" and nothing else. It appeared that the FTC was sympathetic in Carroll Chatham's favor, but the JVC and others were not.

All the trade books and even newspapers printed stories about this "David and Goliath" battle being won by the little guy. Not everyone was pleased with this outcome, however. Some people in the industry, probably the same who started the whole fiasco in the first place, were more upset than before. Maurice Shire, a New York emerald dealer, for example, came out of the woodwork complaining about the new advertisements. "This puts Chatham on a level with God now that he 'creates' emerald." Shire was very vocal and outspoken about Chatham at any public meetings or presentations.

Whoever started the first action against Chatham was busy at work, creating another firestorm for the Chatham organization. The original opponents of "Chatham Cultured Emerald" were not appeased by the new terminology, "Chatham Created Emerald." Not only were they not appeased, but they were also enraged by this new wording to the trade and geared up to challenge my father's emeralds again.

So, the Chatham Created Emerald euphoria was short-lived: just six months later, another bombshell hit.

On Nov. 18, 1960, a notice from the Acting Assistant to General Council for Compliance advised Carroll Chatham et all:

> On November 15, 1960, the Commission rescinded its action of July 27, 1960 wherein it accepted the use of "Chatham Created Emerald" and directed Carroll Chatham to modify the term in conformity with the order to cease and desist.

No reasons were stated in the letter for the action taken by the commission on Nov. 15, 1960. Furthermore, the FTC was now demanding that only the word "synthetic" was acceptable to describe a Chatham Created Emerald. This was, in effect, a duplicate of the original cease and desist order originally started almost a year prior.

They wanted Carroll Chatham to call his emeralds "synthetic emerald" and nothing else would do. It certainly looked like it was the Jewelers Vigilance Committee at work again, spurred on by Fifth Avenue jewelers, but this time the fight was not over how Carroll made or "cultured" his emeralds, but what kind of emeralds they were, a completely different fight.

MID-CONTINENT JEWELER

Editorial

The FTC Should Make Up Its Mind

"The Chatham matter has been reopened and a new complaint has been issued and hearings will be held shortly," reads the report of the recent Board of Directors meeting of the Jewelers Vigilance Committee in this issue of Mid-Continent Jeweler.

Let's hope the Chatham matter gets a better shuffle than it did last time ... when it got virtually no shuffle at all. Maybe the ruling that "synthetic" must be used to describe man-created stones resulted from a desire on the part of the precious stone men to protect the retail jeweler and his customer. Regardless, it also curtailed a lucrative new area of medium-priced jewelry store merchandise not plagued by discount competition and extra volume that the retailer needs badly.

The trouble with the word "synthetic" is that nobody likes it. To almost everyone it means "fake." How would you like someone to describe you as having a synthetic personality? Moreover, if you consult the dictionary, you'll find the definition of "synthetic" there is not at all applicable to what the jewelry trade means by synthetic. In fact, the dictionary's definition is just as applicable to glass.

As far as I am concerned "cultured" connotes something grown by man in a tank under pressure with heat ... whether the FTC thinks so or not. "Man-made" may be more accurately descriptive ... but FTC absolutely forbids Linde, for instance, to use this term. So the producers are left with long, drawn-out scientifically accurate names for their products that would be deadly in advertising copy. Names that would frighten a customer away rather than intrigue him to a jewelry store.

Nor can I understand why these terms would detract from genuine emeralds. If anything the man-created stones will increase the knowledge of and appreciation for the genuine. Besides, 80% of most jewelers' customers would not buy genuine emeralds. They can't afford them. They are as expensive or more so than diamonds. How many genuine ruby and/or genuine emerald neckpieces are sold by the average jeweler each year to someone who wants to spend $20,000? Rather the genuine rubies and emeralds most jewelers sell are small stones in rings. But think of the number of necklaces that could be sold with man-created emeralds for $800 to customers who couldn't afford to pay $4,000.

Let's face it. Most jewelers make their livings off the customer who has from $50 to $200 to spend for something they think looks really nice. It is nice if it's a man-created stone set in a platinum or gold mounting ... especially if the design of the setting includes some diamonds. But that customer doesn't want it called "synthetic." He'd rather buy a necklace with stones of glass and let it go ... or spend the money on some other non-jewelry item.

Most important, the Federal Trade Commission had better make up its mind ... and soon ... about what it's going to permit the man-made stones to be called. If it doesn't, they're going to be referred to by such a myriad of terms that no one ... jeweler or consumer ... will really know what he's buying and/or selling.

Hoyt Hurst

Editor and Publisher.

Hoyt Hurst, editor and publisher of *Mid-Continent Jeweler* magazine wrote an editorial in response to the FTC re-opening the discussion about what to call Carroll Chatham's emeralds. Many sympathized with my father's plight and the industry was very divided about nomenclature: man-made, synthetic or laboratory-created vs. natural. It is only slightly better 70 years later!

This appears at first to be a duplicate of the original complaint regarding "cultured," but the FTC went after what "cultured" meant instead of demanding only the use of the word "synthetic." The FTC doesn't tell you what word to use, it only judges on what you do use. The court got bogged down in semantics rather than what the JVC and others really wanted, "synthetic emerald" and nothing else. It appeared that the FTC was sympathetic in Carroll Chatham's favor, but the JVC and others were not.

Everyone involved on our side agreed to fight it to the end, so everyone headed back to Washington, D.C.

All the same experts, plus many more, were called. Over the next two years, they argued over the definition of "synthetic" and its misuse in the English language. In the first trial, we were bogged down with explaining "cultured" and how it applied to what Carroll Chatham made. Now, we were defining the misconceptions of the word "synthetic," a much easier fight on the surface.

The GIA experts stated that a synthetic stone must be the same in every aspect – the chemical structure, and all physical properties must be essentially identical to their natural counterpart. Chatham's lawyers pointed out all the "synthetic" products in the marketplace that had no resemblance to their natural counterpart, such as nylon, rubber, leather, oil and many more, all legally called "synthetic." Our attorneys even went so far as to suggest that they pull in a few "Joe citizens" off the street to hear their definitions of "synthetic" and measure its resemblance to GIA's definition. This was rejected, of course, because the opposition knew they would lose.

Back and forth they went, and it really tore my father apart and no doubt cost him many years of his life. My father was a scientist,

not a politician or a street fighter, but he was fighting for his life's work.

After an unbelievable amount of time, energy and money was spent, the FTC adjourned and said a decision would be forthcoming. The company was on hold – no advertising, no promotions, no sales.

My father kept how very concerned and exhausted he was very well hidden. He was a very quiet man and kept everything inside, never revealing his innermost thoughts to anyone, I don't think even to my mother. So, for my brother and me, it was life as usual. My mother never went to the hearings in Washington. I was only 16 at the time so had no idea what was going on. To me, life was the same as usual, going to high school in Santa Cruz and surfing.

Finally, a decision came through the mail.

June 25, 1964

Re: Docket 7609, Carroll Chatham, et al…

Gentlemen:

The Commission has reviewed the report of compliance and has concluded, on the assumption that the information submitted is accurate and complete, that the actions set forth therein constitute compliance with the order to cease and desist. The Commission, however, may at any time reconsider, revoke or rescind such approval should it subsequently appear that such information is inaccurate or incomplete, or that actions have been taken in violation of the terms of the order.

Joseph W. Shea

Secretary

FEDERAL TRADE COMMISSION
WASHINGTON 25, D. C.

OFFICE OF THE SECRETARY

JUN 25 1964

Chatham Research Laboratories,
70 - 14th Street,
San Francisco 3, California.

Attention: Carroll F. Chatham.

Re: Chatham Research Laboratories,
et al.,
Docket 7609.

Gentlemen:

The Commission is in receipt of your communication
dated April 23, 1964, which you have filed as a report
showing the manner and form of your compliance with the
order to cease and desist issued on February 28, 1964, in
the above case.

The Commission has reviewed the report of compliance
and has concluded, on the assumption that the information
submitted is accurate and complete, that the actions set
forth therein constitute compliance with the order to
cease and desist. The Commission, however, may at any
time reconsider, revoke or rescind such approval should
it subsequently appear that such information is inaccurate or incomplete, or that actions have been taken in
violation of the terms of the order.

By direction of the Commission.

Joseph W. Shea,
Secretary.

cc: Caesar L. Pitassy, Esquire,
Royall, Koegel and Rogers,
200 Broadway, Park Avenue,
New York 11, N. Y.

Ephraim Jacobs, Esquire,
Hollabaugh & Jacobs,
910 - 17th Street, N.W.,
Washington, D.C. 20006

This June 1964 letter rescinds the "cease and desist" order from FTC and is how the government let my father know that they accepted "created" emerald after four lengthy and expensive years in a Washington, D.C. courthouse. There is no compensation when fighting the federal government, even when you win!

My father had won! Everyone was stunned! Chatham Created Emerald was now permanent. Four years of deliberation. Huge expenses, not to mention the loss in sales because we could not advertise during this whole process. No pay back, no apologies, just a "thanks for coming by folks" sort of letter. Collectively, my father and his partners had spent hundreds of thousands of dollars to prove they were innocent. In federal court, you are presumed guilty and it's your job and financial responsibility to prove otherwise. You win? No compensation, unless you sue the government for it, another lengthy, expensive process you may lose as well. The FTC court system is still run this way, with this added warning: If the FTC even THINKS you're doing something misleading, you are presumed guilty and you must prove otherwise!

I never knew who paid the brunt of those expenses, and they may have been shared equally for all I know. It didn't affect our way of life, which was quite simple. Many people had the misconception, and still do, that Carroll Chatham was immensely wealthy and that these little legal exercises meant nothing to him financially, but I know they did, and it was costly. Yes, it was a write off, but still a huge expense. Back then, it was healthy to have big deductions.

If you made over $100,000 a year in the 1960s, the tax was anywhere from 92 to 96 percent. Add California income tax and you could exceed 100 percent tax. How is that possible? Deductions and write-offs were the name of the game back then. My father wasn't raised that way. Need a car? Pay cash. Buy a house? Pay cash. While most people would have interest deductions and all sorts of write-offs for business equipment and operating costs, Carroll had none. He had zero employees. He built everything in his labs

"Many people had the misconception, and still do, that Carroll Chatham was immensely wealthy and that these little legal exercises meant nothing to him financially, but I know they did, and it was costly."

– Tom Chatham

by hand. He had no outstanding financial obligations, but many people like my father suffered greatly from the tax laws of the time.

Luckily, these legal expenses were 100 percent deductible in the year of expense. It was a small compensation. Instead of paying the IRS we paid the attorneys!

My father eventually hooked up with an accounting firm, Arthur Anderson, one of the big five accounting firms in the U.S. at the time and did get around some of these unbelievable tax rates, that along with the court expenses, were a huge financial drain.

10

Lessons Learned Growing Up

Growing up in my father's home meant doing well in math and chemistry. If you flunked English, OK, or political science, no big deal – but math and chemistry were sacred and must be mastered.

Our living room was often used for business since no one was allowed in the lab. My brother John, (left) who was three years older, and I (at age 3) played with emerald crystals while dad and mom read on the couch for this photo for a 1947 *Fortune* magazine article on my father's work.

My father taught my brother and me a lot of lessons as we were growing up. He might not have said them out loud, but he showed us by example.

He built a small lab in our basement at home, for instance, to show us some of the wonders of chemistry. Once he purchased a 12-foot weather balloon – available for just a dollar at the numerous Army-Navy surplus stores in San Francisco after World War II and filled it with natural gas. My brother and I had no idea what was going to happen or what our father had in mind, but he snaked a long tube connected to a Bunsen burner outlet on the work bench through the lab window to our back yard.

As the balloon filled with the gas, it began to rise, natural gas being lighter than air, until we had this enormous balloon about 12 feet in diameter. It was early evening in winter, so it was just getting dark. While my brother and I held this enormous balloon, my father took about 40 feet of string and dipped it in some kerosene, which is flammable, but not a fast burner. He used a rag to wipe down the full length of the string to get it dry, then, he attached it to the balloon and let it slowly rise to the full 40 feet. We had no idea what was to come, but just this huge balloon over our house was awesome.

To our surprise, my father then lit the end of the string with his lighter and let the balloon float upward. Up and up it went, with this little flame trailing below it, to about 1,000 feet. After about five minutes the flame hit the balloon – not a sound was heard, but a huge flash of a fireball about 20 feet in diameter appeared for about five to 10 seconds. My brother and I were awestruck!

The weather balloon could have come down on someone's home and started a fire, but it didn't. Somebody saw it, however, and fire trucks and police cars were soon prowling around the neighborhood looking for evidence of this UFO. It was spectacular!

We were told never to do this on our own. I have no idea of what the lesson was for, except fun with chemistry, but we learned to respect things that could seriously hurt someone.

Over the course of many years, my father held court in our home's living room. My mom, brother and I often sat in during these meetings, which turned out to be a huge advantage for me and eventually led to my running Chatham Research Laboratories. Of course, being only 4 or 5 years old at the time, I had little to offer, but I did get a sense of the different personalities that came to visit that would serve me well in future endeavors.

Magazine writers, opportunists looking to make a fast buck, top executives from many companies and top gemologists from around the world would visit. Even competitors would call on my father. I would sit and listen and watch. I learned what was an appropriate question and what wasn't, usually from my father who would declare, "That is none of your business." Having sat through so many of these visits I had no trouble talking with people far superior to my knowledge or

experience without feeling inferior or overwhelmed as I grew older.

There was no living off dad in our house, a practice he instilled into me and my brother at a young age. If you want something, go find a job and earn it, and we did. We sold Christmas wreaths during the holidays, and I had a paper route delivering the *San Francisco Chronicle*. I had to get up at 4 a.m., work for two hours delivering papers, then go back home to sleep another hour before school. I also had to collect the money for each house I delivered papers to every month. My paper route job taught me a lot about discipline of purpose and how to deal with people who wouldn't pay the $3.50 per month fee. I think I got to keep a $1 of each customer's monthly payment. Some never would pay, but some would tip, especially around Christmas.

I was learning how to become financially independent, much to the chagrin of my parents, because I would go buy things they would never consider buying, such as a little motorcycle called a Doodlebug for $5. Sure, the engine was blown with a piston rod sticking out the crankcase, but it was a project, bought with my money. Another was a 1934 Ford, five-window, gangster-type car, complete with a purple and black tucked-and-rolled interior for $150. I was 15 years old.

"You can't drive that without a license," my mother yelled at me. "Why did you buy that car?"

Because I could. Houses in our neighborhood were not huge, maybe 2,000 square feet, but they all had big garages underneath. With a little finagling, you could put five cars in one garage. My father built a lab/workshop in ours and my brother and I became pretty good at mechanical things early on with this workshop. Engine building and hot rodding were just coming into vogue during this stage of our lives.

Growing up in my father's home meant doing well in math and chemistry. If you flunked English, OK, or political science, no big deal – but math and chemistry were sacred and must be mastered. I attended high school and college in Santa Cruz, California from 1961 to 1965, cramming all I could into those two precious subjects. After two and a half years, I came to an important conclusion: Carroll Chatham, I was not.

I loved building engines and off-road race cars, and I guess I inherited that can-do spirit from my father and his adeptness at building from scratch and making things work, but I never developed his "love" for chemistry; it never struck a chord with me. I asked my father for a job in late 1965. I loved to work, but I did not like school that much. It seemed pointless and taught me nothing, while my after-school job in an auto body shop, gas station and engine repair shop was the best education I could ever want. I was prepared to either go to United Airlines and sign on as an apprentice mechanic or work for my father. I was 20 years old.

Chatham Research Laboratories never had an employee because my father was not keen on government regulations, especially after the FTC fiascos. Too many forms, rules, taxes, insurance, etc. He said I would be an independent contractor if I was going to work for him. So, first thing I did was study up on "What is an independent contractor?"

The learning and research were invaluable. I even went to the Federal Building in San Francisco and challenged my requirement to pay into Social Security. It had just become mandatory and unlike being an employee, an independent contractor was required to shoulder 100 percent of that cost of about

$300 a year back then, now it's more than $18,000. I studied all the tax rules and regulations. I was informed by the IRS of my obligation to pay Social Security taxes, or join the clergy – who were exempt, but not my choice – so I had to pay.

I found other advantages that also went along with being an independent contractor. I could expense my car, my gas, part of my home as an office and all the tools I bought; all the little things you take for granted and pay for when you're an employee. I also wrote off donations of materials such as tile, clothing, engine parts, etc. because I was filing as self-employed on form "C."

One year I got called into the IRS office to explain my donations. Today you need appraisals from independent sources, but not back then. Anticipating this question because they usually tell you what to bring, I said, "Well, since I donated this stuff there is nothing to show you, but here is a file on what we give away through our business, thousands of dollars in stone samples. That is my personal philosophy, give to others in need." Case closed!

Today, it is a little harder to be an independent contractor, no doubt because of all the loopholes it created. But for many years I was one and came and went when I pleased, had no one telling me what to do, and was on my own – sort of. And my net income grew under my father's tutelage: He was like a walking textbook, but never showed it.

I learned more about math and chemistry from him than any time in college. I know what my father was famous for – growing emeralds – but I had no idea how he did it and I did not learn how it was done for quite some time. I don't think he even understood how good he was, unfortunately.

11

Inventing an Emerald-Growing Furnace

The furnace my father created was an important invention – which has never before been revealed – and it made our production very efficient.

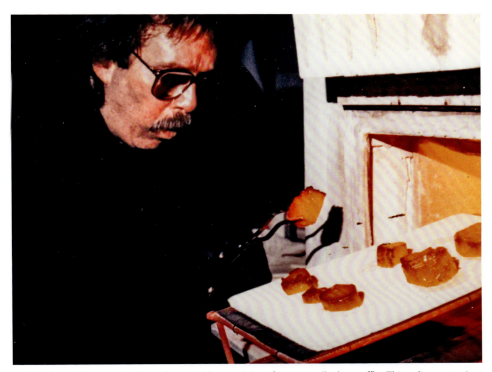

My brother John anneals (heats) emerald crystals in a furnace called a muffle. This relieves strain in the crystals. The temperature is about 900 Celsius.

So, what does one do when starting to work for Chatham Research Labs? Well, you sweep the floors, clean out beakers, learn to weld exotic metals and rebuild furnaces. I did anything that had to get done so my father didn't have to. It was hot and dirty work, but I excelled at it.

I really didn't know all the different work my father did and had perfected over the last 20 years, but I knew there were no stores to buy the equipment he needed, so my father would make it himself. Little by little I learned how to do these things, too. From there, it was a big jump to get involved in the emerald growing part of my father's job.

Growing gemstones is not a simple procedure. It is more than assembling parts, putting chemicals in a pot and turning on the power. My father figured out what happens in the earth to create an emerald and how to duplicate those conditions in the laboratory.

Loading a furnace crucible with flux material:

1. Diatomaceous earth is used on the bottom as an insulator, then a layer of fire bricks with a clay disc is placed on top.

2. The heating element is placed on top of the clay disc.

3. The crucible is placed in the furnace and the flux material is added.

The furnace my father created was an important invention – which has never before been revealed – and it made our production very efficient. It consisted of a simple metal can, about the size of a garbage can, with a layer of diatomaceous earth on the bottom. Diatomaceous earth is an extremely fine talc-like material made up of dead diatoms, a prehistoric bug that lived in the ocean. As these bugs died, layers and layers of them built up on the ocean floor over billions of years. When the ocean retreated and the grounds were exposed to the air, they dried into a hard cake. You can find diatomaceous earth by the mile in the Nevada desert. I had raced off-road vehicles over it and knew that if you slow down too much, you sink in it as if in water and can't swim out. It is ultra-fine like talc, but dense if packed together tightly.

Diatomaceous earth is a superb insulator and was perfect for our furnaces, although it's now considered carcinogenic and safety equipment is used to avoid inhaling it. My first job in learning how to grow emeralds was figuring out how to work with diatomaceous earth. The bottom of our furnace cans had about 10 inches of it that had to be packed down to a firm layer. If you pushed your hands down in it too quickly, they would go straight to the bottom, so you had to gently pat and push it to release the air and slowly firm it up. Of course, by then, you're covered in it.

Fire bricks are placed on top of this firm bed and then a clay disc that we fabricated to be ½-inch thick and 12 inches in diameter was placed on top of the fire bricks. Next on the stack was an element bed my father engineered – using his mathematical expertise and the laws of electricity – to hold the exotic metal heating element, based on how much resistance was present in the metal. It was made from about 20 feet of metal rod, which will go unidentified.

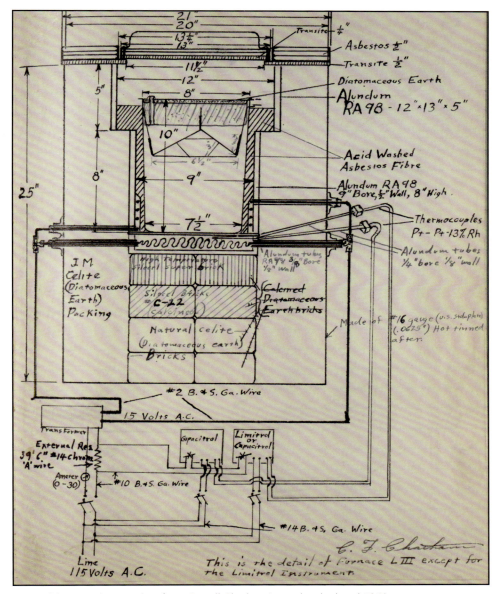

Original furnace design taken from Carroll Chatham's notebook, dated 1940.

A special hand-made form was built to wind the rod back and forth until the element looked like the ones on an electric kitchen oven. It was precisely calculated to reach a certain temperature with a continuous voltage feed. The tails of the rod exited the can to be electrified.

This coiled element was placed in another handmade clay form to keep it from snaking around and shorting out. Another 10-inch clay plate is placed on top of the element and then a cylindrical muffle was put on top of the element plate. We used to buy the clay cylindrical muffle from Norton Ceramics, but at $1,000 each, we soon learned how to build

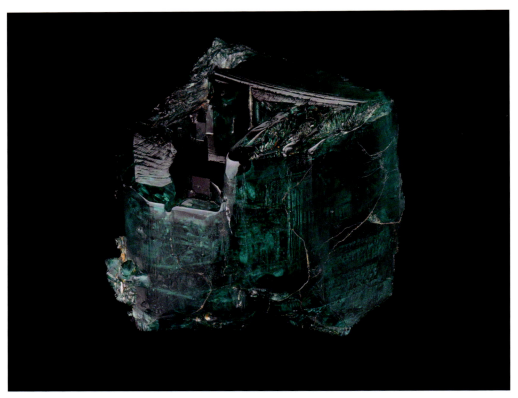

Even when we grew large single crystals – this one is about 800 carats – they had many flaws that required the expertise of well-trained natural emerald cutters to make them into jewelry quality stones. Credit: Orasa Weldon.

those too. A platinum crucible about 10 inches high and eight inches in diameter is placed inside the muffle. The thickness is about .10 inches, like a piece of sheet metal but far more flexible and expensive. The crucible weight was approximately 20 ounces.

Before I came aboard, Engelhard Industries – a huge precious metal refiner and fabricator in New Jersey – used to build beautiful crucibles for us, but too many failed during a production run. If even the smallest amount of "flux" or molten solution inside the crucible reached the electric element due to a pin hole leak, the furnace burned out and all was lost. My father and I visited their factory and found that once the crucible was made, it was spun on a mandrel to make it look pretty (and worth the $5,000 labor fee, no metal cost included). Engelhard sympathized with our problem, but they had no experience with someone who was using such thin metal for so long a period. Their only solution was to make it thicker, but that meant double the cost.

We asked Engelhard to stop spinning our crucibles because when the metal was spun, the potential cracks and/or pin holes were covered up, but not permanently. We received much uglier crucibles, but all the blemishes were visible. Next, we took a page out of my automotive machining background and tried the same method used to detect cracks in crankshafts. This involved painting on an extremely viscous oil that would sneak into cracks as small as 20 millimicrons, wiping off the excess and spraying on a "developer." Then you used a short-wave ultra violet light in the dark so all the cracks would leap out, and you knew where to weld them closed. The welded journals would be reground to the specific bearing size.

We reversed this process with the unspun, not so pretty Engelhard crucibles. We painted the penetrating oil on the inside of the crucible, and then sprayed the outside with the developer. Bright green dots under short wave florescent exposure appeared at any leak, usually on a welded seam. We marked the dots with a black marker, cleaned it up and re-welded the spot. Welding .10-inch material is not an easy trick, however, and we spent hundreds of hours experimenting with how to "float" a puddle of melted platinum without creating a big hole that would require a patch. Bottom line, we rarely had a leak after this discovery and development of this tracing fluid.

Once we figured all of this out, I proposed to my father that we make our own crucibles. At $5,000 each, we spent more than $100,000 a year back then for crucibles from Engelhard. This was labor only and did not include the platinum cost. We made our forms and tools out of nickel metal, which did not react with platinum. We bought the platinum sheet from Engelhard and used our own templates that we made in sheet metal. Since I had already mastered the art of welding thin sheet platinum, this was a logical step and saved a lot of cost.

By 1970, I was turning out a crucible in about an hour, tested and ready to go, saving $5,000 a pop. We never got into refining and producing the metal sheets ourselves – that truly was out of our field of expertise and ability – and

always relied on others, like Engelhard, to provide them. We knew other growers who made their own metal sheet, but it was always too thick and a waste of money. Buying a finished metal sheet of a known thickness and purity was much cheaper and a smarter way to go in the long run. Plus, platinum was getting popular and prices were going up, at the time over $800 per ounce and climbing. Keeping the sheet thin was important, cost-wise.

So, now you have a furnace and a crucible. Next, you need temperature controllers, which we bought off the shelf. They also failed regularly, so we had to take them apart on arrival to get all the metal shavings and garbage out of the insides. This was way before solid state electronics and the temperature controllers were built using tubes, like in a radio, and all had their own personalities. The problem was, nobody had invented anything that had to maintain a consistent temperature over such a long period of time: one year, at plus or minus 1 degree Celsius.

In fact, Corning Glass Works paid us a visit when they heard about our work, and we helped them understand the problems with the mechanical temperature controllers we were all using. We pointed out the sloppiness of the electronics, the dirt that caused hiccups and how we used two controllers per furnace to protect against overruns by a stuck on/off switch (everything was mechanical inside). They were very thankful and treated us to dinner.

We were always helpful if you weren't a competitor and if our emerald process wasn't compromised in anyway. We even taught Engelhard how to make a better crucible because the electronic chip industry was just beginning to develop their own crystal growing experiments.

12

How to Grow Emeralds

It was pure joy to open a furnace once it cooled. We used a flashlight and mirror to see exactly what had grown inside. Were there any anomalies? Did the crystals look clean on the surface or were there tiny stones that, like a cancer, ruined the batch? Notes were taken and all details were recorded, good or bad.

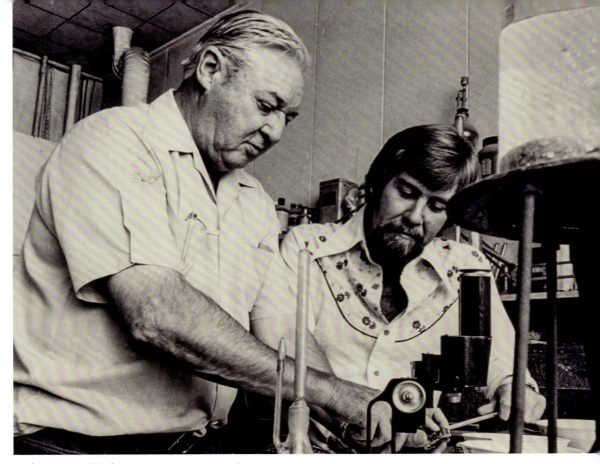

My father, Carroll (left) and I in the shop area of the lab working on a problem. This was about 1967.

It's time to make emeralds. Emerald is composed of beryllium, silica, alumina and chromium, but you cannot simply melt those elements together into a crystal. My father discovered a combination of chemicals that when melted, would dissolve the elements that make up an emerald. This is called a flux. It does not end up in the final emerald crystals but acts as a solvent of sorts that allows the elements to interact. Discovering what elements to use in this flux is a closely guarded secret even today.

It took about two months to set up the chemical base, taking it up to a certain temperature and holding it there. Other elements were added and when small emerald crystals began to form on the top of the melt, it was time to add the seeds from which the emeralds would grow from.

At first, Carroll used natural beryl seeds for this step, but due to quality and the water present in their natural structure, we soon turned to our

John Chatham scoops out the molten flux from an emerald furnace. The temperature is about 900 Celsius inside the platinum crucible, which is too hot and dangerous to wear more than shorts to avoid injury in case of an accident.

own production, sacrificing some crystals we had grown to use as seed crystals for the new emerald to grow from. (We still have some of those emerald crystals with a natural beryl seed in our display at our corporate headquarters in San Marcos, CA.) The seeds were heated up to about 900 Celsius and very carefully floated on the surface of the melt. The clay lid of the furnace was lowered, and then another lid made of asbestos capped off the top of the furnace. The handle was then removed. No one touched that furnace for 10 months. This entire set up took two months to assemble for a total of 12 months.

As you can imagine, the furnace room was hot – a constant 90 degrees – but it was far cooler in that room than one would think with 20 furnaces running at 1,100 Celsius. This was the beauty of my father's furnace design. You could touch the outside skin of a furnace that contained 1,100 Celsius molten materials – only 10 inches away from your hand – and not get burnt. The insulation was fantastic and saved a lot of energy.

We only wore shorts – no shirts – when we worked on a furnace. We had two reasons for this protocol: First, if anything splashed on your clothing it would catch fire immediately and be hard to get off before you were badly burned. Second, we didn't wear shirts because if there was a splash, the moisture on your skin would boil and burn but there would be no cloth to stick to the skin, which would be much more damaging.

We were extremely careful around these furnaces and never had an accident in more than 50 years. Our biggest fear was earthquakes, floods and burglaries. Having hundreds of ounces of platinum was a hard secret to keep and if some idiot came in and tried to empty a furnace, they would not have lived through it – nor would the building.

After a total of 12 months, we knew the available food/chemical supply used to grow the crystals was about exhausted inside the crucible. If you let it run longer, the crystals would start to dissolve and grow someplace else, usually not where you wanted them. As the crystals grew, sometimes an inch long, their physical size and position would change the natural convection of the molten chemicals, creating hot spots (relatively speaking) and would re-channel growth material in all the wrong places, so 12 months was the cut off.

The exterior metal handle was reattached to the furnace to open it. Only a small hole of about one inch was visible because all the emerald crystals had grown together. We used a device called a pipette made of platinum to suck up the remaining 1,100 Celsius molten liquid flux surrounding the emerald crystals.

So, a rhythm was established: suck up the

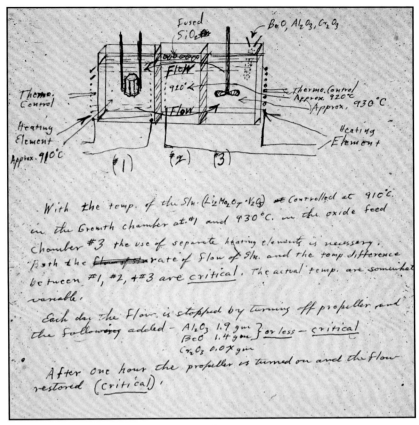

Carroll's rough notes on making emerald from 1936.

molten liquid until the chamber halfway up the tube was filled, bite the tube with your mouth to hold the liquid up, and quickly release the liquid into a water-cooled device we also invented to contain the hot molten flux. The liquid material immediately solidified. Imagine a straw in water; when you cap the end the straw, it holds the liquid until you remove your finger.

It sounds scary, but the pipette we used had a chamber halfway up the tube that would hold about six ounces of fluid, so you could see how high the liquid was coming up the pipette because it would become red hot. You could not hesitate because if the liquid cooled, it solidified in the chamber.

This procedure was performed maybe 20 times, until the level of molten flux was well below the crystals. We never had a misstep because it was done with the utmost care: one drop of liquid down that tube from spit and you are history – or one bad swing with a full load of molten liquid in the chamber and who knows what would catch on fire. We never had an accident. One quickly learns to respect molten flux at 1,000+ Celsius.

Then the furnace was closed again and the power turned off. It took about two days for the furnace and the crystals in it to completely cool down. This was an annealing step to avoid causing strain and cracks in the crystals.

It was pure joy to open a furnace once it cooled. We used a flashlight and mirror to see exactly what had grown inside. Were there

The cover of *The Bay Viewer*, 1970.

any anomalies? Did the crystals look clean on the surface or were there tiny stones that, like a cancer, ruined the batch? Notes were taken and all details were recorded, good or bad.

Next, the furnace was taken apart all the way down to the fire bricks on the diatomaceous earth and if there were any tiny leaks in the crucible, everything was thrown out. The crucible was then placed on a ceramic table and cut away, reveling a large circle of emerald crystals all grown together.

The circle of emeralds was weighed (usually about 10,000 carats), broken up with a hammer and put in a hot water bucket to clean off any remaining flux. Next came hours of sawing the seeds off the crystals, trying to maximize the greatest yield of the best quality new crystal growth. The crucible walls were also covered in small spontaneously grown crystals. We called them "clusters" and sold them to be used as un-faceted gems in jewelry – sort of like druzy quartz is mounted today.

Every furnace run was different: some furnaces consistently produced the best quality emerald crystals, while others always did poorly. There was absolutely no reason for these differences in crystal growth since every furnace was totally rebuilt after every run. The transformers for controlling power input were all identical by the electronic measurements taken. The temperature controllers did vary from furnace to furnace, but even once we switched to digital controls, the anomalies continued.

Growing crystals, even under controlled laboratory conditions, is an art one never truly masters. Just like crystals that formed in the earth's crust, we were always amazed at the individuality of each run. I suppose this is like making a fine wine, where everything from soil conditions, climate and finally the expertise of the vintner is put to the test; some are great, some not so great.

There are so many variables in both earth-grown crystals and laboratory-grown crystals; it precludes establishing an absolute formula or receipt. Why do inclusions form? Why is one crystal gem quality while one next to it is cloudy? This was probably the most frustrating aspect of growing any type of gemstone crystal. My father once remarked; if this business model waited to achieve perfection, we would never be in business. Just like in mining natural gemstones, we took the good with the bad and developed markets for all qualities of stone.

My father gave me one furnace I built, No. 26 (out of 65 furnaces), as a reward. It was the worst producer in the lab, no matter what I did, year after year. It hardly even paid for itself. We finally took it offline and left the space it had occupied bare. Actually, no I take that back, we put a garbage can there! To this day, I have no idea what was wrong with No. 26 and it continues to bother me to this day.

World's only emerald grower ...where but in San Francisco!

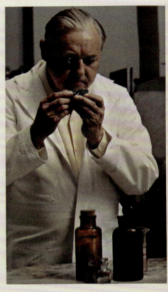

 When Mrs. Carroll F. Chatham wants a new emerald, she doesn't reach for cash or a credit card. She asks her husband to grow her a stone. Which he can—and does—in his San Francisco plant.

The precious baubles Chatham creates actually are "grown." The principle on which Chatham creates his shining treasures can be compared to the method used to make rock candy. Just dip a string into super-saturated sugar syrup. Crystals form, and on them more crystals gather. The new crystals that form have the same composition as the "starter" crystals. Result: Rock candy. To make emeralds, start with tiny seeds of natural, rough emeralds. Over a period of a year, crystals will build up in this same rock candy fashion. Result: Created stones of rare beauty—chemically and physically similar to the finest Colombian emeralds.

Only firing to a red heat or examination under a microscope can reveal which is man-made and which is mined. A mined emerald will crack or completely disintegrate under such heat; the Chatham-created emerald will not. X-ray examination proves the molecular structures identical, but the laboratory stones lack the traces of impurities which nature introduces when she produces emeralds. Natural or man-made, no two stones are alike.

It all sounds so easy. But before you hop out to the kitchen to run up a few necklaces of your own, consider the years Carroll Chatham spent in perfecting his process.

He started in 1928, while he was still in grammar school, working at home with home-made equipment. By the time he finished high school, he was producing emerald crystals, but they were as worthless as gravel. A degree in chemistry and years of night work followed.

Then perseverance paid off, in 1939. The inventor sent his first created emerald of quality to the Smithsonian Institution. Another of his early successes shines in his wife's engagement ring.

Current production is small—roughly, 300,000 carats of rough material, not all of which ends up as jewelry. It is possible to grow a green-eyed monster as large as 2,000 carats, but those over twenty carats lack the fine quality jewelers demand.

Chatham-created emeralds sell at retail from $70 to $400 a carat, depending on weight, quality, and color. More costly than diamonds, fine natural emeralds are valued at many times the price of comparable Chatham crystals. Chatham-created emeralds are cut by the same skilled gem-cutters who work on gemstones from the mine.

Anyone may feel free to duplicate Chatham's feat; his process is not patented. It's available to anyone who can discover its secrets. But there will be no helpful hints from the original inventor; Chatham talks about the process neither in his sleep nor in his waking hours. Equally secret is Chatham's process for growing ruby crystals.

The next Chatham goal: Synthetic diamonds for industry. Who knows? San Francisco may yet boast of having the world's only diamond farm.

Carroll F. Chatham takes a close look at a rough crystal recently harvested. It is extremely difficult to photograph both natural emeralds and Chatham-created emeralds. Ironically, only fake stones photograph green. True gems usually show up brownish-red in a printed picture, and only a photo-laboratory can re-color them green as seen here.

So many articles were written over the years – all for free, which was a major help in our company's growth. Inside pages of the 1970 *Bay Viewer* article.

13

Upsetting the Profit-Sharing Apple Cart

"Dad," I said one day, as I wrote a check to the government for more than $200,000 in income taxes at the end of the year, "This doesn't feel right. You have one customer, Dan Mayers, who doesn't even touch the stones we ship to New York. Where do they end up and how much does he make on the deal? It seems like he is making more money off our emeralds than we are."

– Tom Chatham

Within five years I was fully in charge of running the labs, grading the stones, and cutting the rough for seeds. I was also doing all the welding and fabrication for the furnaces – and putting my two cents in whenever a problem came up – and come up they did.

For instance, we had dozens of furnace runs where non-emerald related crystals would grow. It was probably caused by a contaminate, maybe a few parts per million, in one of our base chemicals. My motto was, "when in doubt, throw it out!" Buy new chemicals. My father, on the other hand, wanted to figure out why it was happening.

One year we were plagued with phenacite crystals forming inside our emerald crystals as they grew. This is a beryllium silicate compound. After careful and painstaking analysis, my father figured out the cure: a simple rebalancing of certain chemicals. To save time and costs, some furnace runs were sacrificed six months early to confirm his diagnosis. His solution was more scientifically correct, but it cost us years of time and hundreds of thousands of dollars in lost production.

I liked to work, and I finally got to know my father from working so closely with him every day. I was grateful for this since I grew up without him because of his dedication to work. He would sleep in until 9 a.m. when I was already at school, and he came home around 10 at night, when I was already in bed. The only time we saw each other much was when I was in trouble for something, so it was not a great traditional father-son relationship.

My father also favored my brother John, who was the oldest and an excellent student. John was sent to military school and got straight A's. I pleaded to go, too, but I guess my parents thought it was a waste of time because I was not a great student. Going to work with my father after only two years of college and excelling at every task made us much closer on all levels, both in our family and business relationship.

My father was no doubt a chemistry genius, but accounting was not his forte. Oh, he could do books and file his own taxes, but didn't have that drive to understand all the economics of running a business.

"Dad," I said one day, as I wrote a check to the government for more than $200,000 in income taxes at the end of the year, "this doesn't feel right. You have one customer, Dan Mayers, who doesn't even touch the stones we ship to New York. Where do they end up and how much

My father and me, around 1967, in the shop area of the original San Francisco laboratory. A new laboratory dedicated to growing flux ruby and sapphires was built in south San Francisco, about 15 miles away, soon after.

does he make on the deal? It seems like he is making more money off our emeralds than we are."

This didn't make good sense – and to top it off, our tax bracket in the 1960s was about 92 percent. We were making money – good money – but Dan Mayers was making more and doing nothing for it.

My father didn't really say anything, but that was the beginning of the end for Dan. I had planted the idea and put it in motion, and my father knew I was right.

The next time Dan visited us, I casually brought up the subject of marketing and asked where our products sell. He generalized an explanation of the marketplace like it was some mysterious dark hole, while I wanted to know what was selling and in general, where. I wasn't looking for names or companies, just general stuff we should know about. Is the U.S. our only market? Do we sell in other countries? If so, which ones? He never answered but would go off on some tangent of personal accomplishment in some other area.

This was the problem as I saw it: Dan agreed to buy 100 percent of our production and paid in advance, but we shipped directly to Ed Coyne in New York and Dan did nothing but mark up the bill! It didn't feel equitable. My father didn't comment but I could tell he was getting the sense of my thoughts and concerns. His silence told me a lot. Except for finding Ed Coyne, Dan did not contribute anything to our continued or future success.

In hindsight, I can see it was easier for my father to run it this way, but, the deal started to stink and could put us in a dangerous position if anything happened to Dan. We wouldn't know what to do with our stones.

During visits to New York to see Ed Coyne,

my questions continued, such as "Oh, by the way, Ed, how much does Dan make on this deal? How is it split, based on your sales or a flat percentage?"

"Dan adds 30 percent across the board on Chatham prices," he told me. I could tell he wanted me to know this because he didn't like Dan Mayers either and getting rid of him could only help his situation.

Once we got back home, the conversations continued.

"Dad, 30 percent for doing what?" I asked. I didn't realize the dilemma I was pushing him into.

Letters went back and forth over this between my father and Dan. My father was very comfortable in the deal, but knew I was right. I didn't really realize what a can of worms I was opening, and my father could have told me to back off anytime he wanted. I would have complied. But he didn't. I was bringing him problems, but not solutions and now he too was questioning the relationship with Dan and his 30 percent markup.

I kept politely pressing Dan in another visit, sometime in 1969 or 1970. Dan thought my older brother John, who was also just starting to work in the labs, was the heir apparent and took him under his wing and set him up in Europe to "learn" the business. The only thing my older brother John learned, however, was to spend money and have a good time on Dan's payroll. That lasted about a year.

I started to wonder how I fit in. Why wasn't I going to Europe? I knew it was for several reasons: I was married, had a young son, and was also becoming indispensable in our two labs. This was not because I suddenly became brilliant, but because I was family and could be trusted. I was becoming a master at doing

everything in the labs and our production was way up and so was quality and gross income. Plus, I was well compensated and had no desire to go sit in Paris and sip cappuccinos. But I still felt slighted. Maybe Dan was trying to elevate John to a superior position because I was a thorn in his side? Well, it didn't work out that way.

At yet another dinner with Dan in a fancy restaurant in San Francisco, he called me aside for a little chat.

"Tom, I think you are upsetting your father with all your questions. You should tone it down a little," Dan said.

"Dan, if my father wanted me to tone it down, he would tell me," I said.

"OK, Tom, here is what we can do," Dan said. "Why don't I set up an account for you over in Geneva, a little nest egg, that will grow every year, and we can learn to 'work' with each other."

"What? Did you just offer me money to shut up and quit rocking the boat?" I asked.

"It would help, Tom," he said.

"Dan, you have just made a fatal mistake in your judgment of my character. If you think for even a second that I would do anything that could hurt my father, you're crazy," I told him. "You cannot buy me or my silence...and I will tell my father about your offer as well."

"I doubt you will do that, Tom," he responded.

"Watch me," I said as I got up, went to the other end of the table, called my father aside and told him about the bribe that was just offered me. He looked at Dan and didn't say another word, but I could feel the hole Dan had been digging get deeper, and deeper.

This old roll top desk from my grandfather's office at the lumber yard was used daily by my father and me to record extensive handwritten notes on the progression of every furnace run and then the outcome of each run. Some runs were great, some were worthless!

Days later, my father wrote Dan a letter informing him of a 10 percent hike in prices that he could not pass on to Ed Coyne of Created Gemstones in New York. On top of that, he had notified Dan that his markup to New York was to be lowered to 10 percent, effectively leaving his commission at only 10 percent after selling to Ed Coyne at Created Gemstones, Inc. The extra 20 percent this produced was to be split between Ed Coyne and Chatham Research Labs. Dan was not doing anything of value to justify a 30 percent markup, in my father's opinion. So, in effect, we raised our prices 10 percent to Dan, and Dan had to lower his prices to Coyne, thus leaving him with just a 10 percent commission.

Dan was livid and let my father know it in a telegram: "You have no idea what you're doing, Carroll. New York will eat you alive! You cannot control these people. Only I can but I will not work for 10 percent." The telegram was never answered by my father.

The next week, my father and I got on a plane to New York to see Ed Coyne at Created Gemstones to explain our new relationship and the end of Dan's partnership with our business. My father proposed that we sell to Ed Coyne directly, eliminating Dan and his 30 percent markup. Ed was elated! He finally got the monkey off his back. We had no idea there was so much animosity that had built up over many years between Dan and Ed. We agreed on the commission split, the payment terms and the grading system. We also agreed on advertising and having approval rights, still stinging from the FTC trials.

The Ipekdjian Brothers were also pleased because it increased their cut of the business that Ed Coyne created. They were like silent partners because of their natural diamond business. I assume they didn't think the two "mixed" well politically in the trade and I rarely ever saw them.

Finally, we were going to find out who Chatham Created Emerald really was!

14

Acceptance in the Trade

This was a very weird situation. Everybody knew our name, Chatham, as in Chatham Created Emeralds, but we knew nothing about the jewelry industry. The credentials to join the AGS required at least one employee to be a graduate gemologist from GIA. That didn't get you in the club, just the front door. You still had to be invited, your company reviewed for its position in the industry and your product had to reflect the highest professional standards of the group.

The courts may have given Chatham Created Emerald its blessing, but certain sectors of the jewelry industry did not, and some never would.

The Jewelers Vigilance Committee was still adamant about using the term "synthetic" to describe our gemstones, but they had no teeth to bite with. What they did have was influence within the trade and trade organizations. When the subject of nomenclature came up for discussion, probably because the FTC likes to review trade regulations every 10 years or so, the fights began all over again, but this time I was on the inside fighting and had the power of a legal precedent behind me with the term "created" forever branded on our product.

I was even asked by some to copyright the term "created" and bar anyone else's use of it, but I refused. My father and I felt that if another company like Gilson or Union Carbide had the same product, they should enjoy the same rights. This was not all nobility on our part, it was also a smarter marketing move so the public would become more aware of "created gemstones" that fit a certain criterion: identical to natural.

I became more active in the politics of the industry around 1970 and joined the local chapter of the American Gem Society (AGS) near San Francisco. This was a brand-new world for me. I was a "rock" star because of my father's work but didn't know a damn thing about gemology. About 50 people would meet for dinner to hear a speaker on some area of interest to the group. I remember vividly how at every meeting, each person would have to stand up, announce their name and what they did or their store's location. I was petrified but did it.

Although I was only 24-25 at the time, I was given credit for my father's 30 years of fame. I felt awkward and phony, and it didn't help that in this group of jewelers was a real rockstar, Arthur Gleim, who was a former chairman of the Gemological Institute of America (GIA) board of governors and former president of AGS. He was a very influential person and extremely nice to me. He knew what my father went through, and I think he sympathized over what had happened with the FTC. He certainly was a gentleman and a friend I could turn to.

One night he asked me if I was an AGS member and I said no, I just attended on a guest basis. "Tom, I think you should take care of that," he told me.

This was a very weird situation. Everybody knew our name, Chatham, as in Chatham Created Emeralds, but we knew nothing about the jewelry industry. The credentials to join the AGS required at least one employee to be a graduate gemologist from GIA.

That didn't get you in the club, just the front door. You still had to be invited, your company reviewed for its position in the industry and your product had to reflect the highest professional standards of the group. If you sold cheap knick knacks and plastic dinnerware, you were not welcome. It was a little snobbish, I agree, but there were a lot of low-end jewelers out there not living up to the honesty and integrity the group wanted to project. It was the elite of the industry.

Great, I thought to myself, what the hell do I do now? I can't just ignore what Mr. Gleim suggested, but I couldn't take six months off to go to GIA in Santa Monica, CA, to earn this diploma.

I learned that GIA had taught one-week classes each on gem identification and diamond grading all over the U.S., so I signed up. This is where I meant Mike Allbritton, my teacher for two weeks. I don't know who was more nervous, me or him. Here I am, the son of a near-legend, sitting in his front row probably criticuing every word out of his mouth, but I only knew about what my father did in his labs. I was not very familiar with other gemstones or the gemology behind them. Chemistry is very different than gemology.

I took to gemology the way my father took to chemistry. I loved it and absorbed everything Allbritton taught me. Since everyone had day jobs, classes started around 5 p.m. and ended at 9, then many of us would go out to dinner or a bar just for drinks. It was an enjoyable two weeks, to say the least. I bought all the books I could on the subject and almost

memorized Richard T. Liddicoat's "Handbook of Gem Identification." It was probably one of the first times I really studied a subject, versus just memorizing something for a test. It felt very natural, and I absorbed it.

In six months, I felt ready, so I called Al Woodill, the president of AGS, who I knew from previous annual AGS Conclaves (education and networking events for members that feature industry experts who teach gemology, marketing and leadership). Non-members could attend if invited, but a supplier like me was limited in what they could say or do at these gatherings. It was a lot of meet, greet and "let me buy you a drink" sort of interactions.

"Al, I want to become an AGS supplier, and I know I need to be a GIA graduate to be one, so is there any way I could challenge the course and take the test to be a graduate gemologist?" Al Woodill, like everyone else in the world, thought I was an expert already because my last name was Chatham, but he stuck to the rules and set me up with a gem identification test and a proctor in my area.

The GIA 20-stone test is the subject of agony and glory in the gem and jewelry industry. You must correctly identify all 20 stones to pass and become a graduate gemologist. Some pass on the first try, but many others need to take it multiple times. The hardest stone for most to identify for decades was the Chatham flux ruby!

I got stuck on one of the stones. I was thinking it was wulfenite, an obscure stone no one ever bumps into, but thought maybe they really wanted to put me to the test. I asked my proctor about wulfenite, and he told me: "Tom, we want you to pass, we are not going to throw in some off-the-wall stone nobody will ever see. Go back and do your tests again." So, I did and realized that I had made

{75}

April 19, 1980

To the Chairman
Suppliers Committee
<u>American Gem Society</u>

We, the undersigned, view with concern what we see to be an erosion of the A.G.S.'s traditional emphasis on quality. We believe it is the unswerving dedication to quality that alone gives meaning to the A.G.S. and its membership. The A.G.S.'s staff and its officers should be steadfastly devoted to promulgating - and living - this basic concept: the best jewelers, the best jewelry. Every advertisement, every brochure, every public utterance should reflect the concept of quality...in the highest degree. And, certainly, at its annual Conclave, this concept should be the keynote!

Yet, for the third consecutive Conclave, credibility is being given to synthetics as a jewelry store item...as an A.G.S. jewelry store item, at that.

Once and for all, we, the undersigned, would like to cut through the confusion concerning the place of gemstone substitutes. Not one of us denies that they will be sold in large numbers, world-wide. There <u>is</u> a place for them: gift shops, specialty stores, department stores. We, the undersigned, however, categorically deny them a place of prominence in the truly fine jewelry store. Tiffany's has never sold a synthetic. Period!

We do not fear the impact of synthetics on our individual businesses. We do deplore the credibility given them by a Society with which we are associated. Even though we are supplier members, we believe we can offer positive advice to the Society. We believe it is needed. The subject of synthetics has arisen either because of thoughtlessness, or lack of professionalism - or both.

Similarly, this thoughtlessness has let to other recent happenings that also affect adversely the basic relationship of the A.G.S. with fine jewelry and its suppliers:

1. A statement in Guilds that: With ownership of the Colormaster, "it is not necessary for the retailer to carry a very large inventory of colored stones".

2. Allowing A.G.S. logos or titles to be used in conjunction with product endorsement.

3. The sponsorship by A.G.S. of travel tours to foreign competitors.

A letter from members of the American Gem Society's governing board protesting Tom Chatham's invitation to speak before its conclave. They believed it was not appropriate to give "credibility" to "synthetics" or "gemstone substitutes" at the event. At the conclusion of my talk, one of the authors

Richard T. Liddicoat, president of GIA, and me at the Chatham booth in Tucson, Arizona in the late 1980s. My father and I had a long and enriching friendship with RTL, as he was known to many, extending into many decades. I believed in supporting the work and research conducted by GIA and donated thousands of stones and many dollars to the Institute for its research.

the classic mistake of trying to shove a round peg into a square hole. I don't remember what that stone was, but it was not wulfenite for sure!

I passed the test and became a graduate gemologist, a lifetime member of the GIA Alumni Association and an official AGS supplier. Since then, I have learned a lot more about gemology and know that one of the most important things to do when identifying gemstones is to figure out what the stone can't be from certain physical tests. This usually narrows it down from about 200 species of stones to maybe five.

I soon rose in the ranks of the AGS and eventually became president of my local chapter. I think Art Gleim was more pleased than me. I also became more active in the national make-up of the AGS as a supplier. One of the speaker committee members, Phil Minsky, asked me to give a speech at the next Conclave in Dallas, Texas in the mid-1980s. This was when GIA and AGS were joined at the hip and did Conclaves together, so the audiences were sizable. It would be in front of the whole assembly, about 1,500 people at the time.

I was scared beyond my worst nightmares. They wanted me to speak about marketing Chatham Created Emeralds as an AGS retail jeweler. "Just 15 minutes or so Tom, we will have three other speakers," Phil told me. So, I wrote a short speech and practiced and practiced in front of a mirror. As nervous as I was, I could not fail at this task, and I knew it was an important hurdle to overcome with long-term benefits for our business. I read my speech word-for-word – something I learned to never to do in the future because it sounded so rehearsed – and opened the floor to any questions.

Microphones were spread throughout the audience floor. A gem dealer I knew and had sparred with before over nomenclature, walked up to a microphone. I don't have a recording of it, only vivid memories, but this was very close to what he said:

I spoke twice a year to GIA's student body in residence at both the Santa Monica and Carlsbad campuses for 30 years. Every student received one of these single crystals at the end of each presentation. The crystals are a byproduct in the Chatham process but tell a story in their natural hexagonal formation. Approximately 20,000 crystals were given away over the years.

"I am flabbergasted!" he announced. "Who in their right mind would allow this outrageous crook to speak in front of such a noble crowd? His family runs the most misleading and crooked company in the industry."

He did not have a question; he just went on and on about how bad my family history was. There was silence when he was done – a long silence. You could have heard a pin drop!

One thing I had learned about public speaking is that you don't get in a pissing match with an audience member in a large crowd … you can't win. The other is, if you're petrified and nervous, the least said the better!

So finally, I simply said, "Thank you for your opinion" and sat down.

The audience responded with overwhelming applause, some even standing up. Phil Minsky, who was moderating, later thanked me for my response.

"Tom, I don't know how you did it, but thanks," he said. "I am so sorry this happened. How you kept your cool is beyond me."

It turned out to be a big mistake on this outspoken man's part. He sounded like an idiot attacking me, my family and many in the audience who didn't even know who he was. He was chastised for his actions, and I think even asked to leave the AGS. The organization was embarrassed, and a lot of members came up to me to apologize for his actions. I am sure I made a few new customers that day.

The man's rants continued for years while he was also taking the position that oiling an emerald to hide flaws was just replacing natural fluids. In my 50-plus years as a professional in this industry, I have never once heard of an emerald mined in an oil field and I know for a fact that injecting natural emerald with oil is an unnatural act. I later heard he got taken on one of our stones for a tidy sum in Colombia and went to bed for a few weeks in depression. I could not confirm this rumor, but it came out of GIA's identification of the stone. He was another one of those old school dealers who depended on their "gut feelings" when buying stones, so it could easily be true that he made a mistake.

Anyway, this was quite a way for me to start my public speaking experience! My then-teacher and now longtime friend, Mike Allbritton, asked me to speak to the GIA student body resident class to share my side of what Chatham Created Gems were and were not. Again, I was scared as hell, but had to accept. There were 350 students and teachers expecting me to speak about how we grew emerald and rubies and the history of my father's work.

I repeated this experience every six months for 30 years. It was always well-received, and

{78}

Acceptance in the trade: Despite a few outspoken individuals, my father and his emeralds were well received and honored at many gatherings among his peers. My father and I were speakers at the first GIA Symposium: Carroll's topic was "Little Known Facts in the Art of Growing Gem Crystals," and mine was "Marketing of Created Gemstones." Even Dr. Edward Gübelin, head of the Gübelin Gemological Laboratory in Switzerland, had to admit his original opinion of Carroll Chatham's work was wrong and deserved its place in history. From left: Carroll Chatham, Tom Chatham, Gale Johnson, Dr. Edward Gübelin, Robert Crowningshield and Robert Kane at the first GIA Symposium in 1982. Credit: GIA.

the students, who would attend voluntarily, seemed appreciative of my efforts. Each received a single emerald crystal attached to a Chatham sales brochure. Those stones are still held by thousands of students who will stop me at a trade show and say, "I heard you speak at GIA in 1986 and still have that crystal ... thanks for the effort!" I figure I have spoken to about 20,000 GIA students over the years, not counting the numerous talks I gave around the world to other gemological groups.

I clearly got over my stage fright and now thoroughly enjoy public speaking. I learned a few lessons along the way: First, you must know your subject inside and out. Second, you should only use brief outline notes, never read a speech. And finally, keep it to about 50 minutes, plus enough time for a Q&A, because people can only sit still – for even the most gifted speaker – for no longer than that.

After 30-plus years, GIA stopped asking me to speak and I never knew why. Maybe it looked too commercial, maybe my competitors thought it was unfair and wanted equal time, which GIA could not do for everyone. Regardless, I reached thousands of future customers over the years and made many friends, both inside GIA and with its graduates. GIA videotaped all my talks so I suppose if any student wants to hear my story, they could see it from the library archives.

I was a major contributor in both stones and money to GIA over the years, but by the time I stopped speaking, GIA was reaping in hundreds of millions of dollars grading diamonds and identifying gemstones from all over the world. The diamond business had become a paper commodity and that paper had better have a GIA logo on it to sell at top dollar.

15

Staying Ahead of the Competition

Having a competitor did not concern us in the least. Many people think being the only producer in the world would make business easy and very profitable. Well, maybe if you make iPhones – not created emeralds. When a Chatham stone was sold, it was because we paid to advertise and market it.

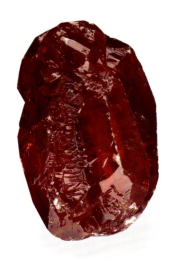

By the early 1970s, I had talked my father into adding another building exclusively for growing ruby, which he first started investigating in the late 1950s. That same year, we also doubled the production of emerald. I could just feel we were on the right track and if we didn't have enough of the product, others were going to overtake us.

Pierre Gilson of France, for instance, had a lifelong passion to do what my father did, grow emerald. He was in the ceramic business with a huge factory in St. Omer, France and made bricks of every type. Because of this, he understood chemistry and high temperature work. He analyzed the trapped flux in some of our stones and was able to guess what our flux chemicals were made up of. He didn't quite get the furnace design right, but he made an excellent emerald product. What he accomplished was nothing short of amazing.

Gilson came to visit my father when he first made his announcements to the trade of his successes in growing emerald in the mid 1960s. He was received in our home, not the lab of course, so I got to meet our first competitor when I was about 18 years old.

He and my father hit it off well and were very similar in temperament. They were of equal age, both born in 1914. Gilson got his inspiration to grow emerald after reading a *Fortune* magazine story about Carroll Chatham in 1947! We became good friends with the entire Gilson family over the years, and he always stopped by on his swings through the United States and we would all go to dinner. We were in many of the same trade shows in the U. S. and Gilson had a good product, but no real representation in the U.S. It was all a game of who you knew, not just what you sold.

I know many aspects of Gilson's business because I bought the intellectual property from his ex-employees 10 years after he sold his company to a Japanese firm, Earth Chemical. France had gone socialist and had some strict rules about selling technology to a foreign entity, as does the United States. The French government fined Gilson and made him persona non grata in France and he had to leave the country, including his big house and property. He could never go home and lived out his life in Switzerland, where he died in 2002 at the age of 88.

Gilson's employees had called me on the phone trying to sell inside information right after the company was sold, but I turned them down for several reasons. I thought it was illegal to do so and unethical. "If you want something for what you know or did, talk to the Japanese owners. I can't help you," I said. Once 10 years had passed, I figured the statute of limitations had long run out and decided to buy the intellectual property.

Having a competitor did not concern us in the least. Many people think being the only producer in the world would make business easy and very profitable. Well, maybe if you make iPhones – not created emeralds. When a Chatham stone was sold, it was because we paid to advertise and market it. When Gilson came online and began advertising, it complimented our product, so the market grew. Then Union Carbide's crystal growing division, the Linde Company, came out with their hydrothermal emerald, which also added to consumer awareness. The market grew exponentially as more competitors came online.

Because of this, I have always had an open-door policy welcoming potential competitors, sellers, aspiring chemists and even goof balls.

Some were suspicious at first: "Mr. Chatham, do I need my attorney present?"

"No, this is not an inquisition; this is a friendly visit to welcome you into the family," I would tell them.

I go out of my way, all over the world, just to meet new producers of gemstones. I also share the past mistakes we made that could hurt us all if repeated. I explain the FTC battles we went through as well as the anti-created gem faction that still exists today in the natural gem market. I also mention some of the people to avoid as being a little unscrupulous. We never discussed pricing or pricing philosophies, and of course production methods were off limits. We generally became good friends, all with a common interest: to sell stones and be accepted.

The last one to come knocking at our door was Truhart Brown from Austin, Texas. He had invented a process for growing flux ruby that was very good and had distinctly different inclusions than Chatham. I remember my father joking with Truhart Brown and his wife Nancy, in our living room, "What are you going to call your product, Brown-Created Ruby?"

No, it was called Kashan Created Ruby and did very well for several years. They took the tack that I advised all to avoid: "Our ruby is so good it will fool even the experts." It did for a while, but soon, enough products got out that if you studied your gemology, it was a straight-forward identification. As usual, everyone did not pay attention and many got burned in Thailand with the Kashan ruby, which did not sit well with the trade.

Kashan ruby, like Chatham rubies, were identical to natural so extreme caution had to be used when buying anything being called ruby, especially in Bangkok or out in the bush. I have discovered over the years that if the trade is afraid of your product, they won't sell it. And if the trade doesn't like the way you market your product, they will shun it like the plague.

One inventor who didn't like my buddy system was Judith Osmer, creator of the Ramaura Ruby.

My father grew some specialty crystals for Hughes Aircraft and other materials research groups on a contract basis in the 1950s. Some of these experiments took years to complete. One memorable contract took three years to accomplish and the people who ordered it didn't even work for the company anymore, but they still had to pay. These researchers were probably working on the first lasers.

Osmer was on one of the research teams

A large ruby from the Athens Lab. Credit: Orasa Weldon.

experimenting with my father's stones and became intrigued with the thought of making rubies on her own. When she left Hughes Aircraft or one of the other aerospace companies she worked for, she contacted me and wanted to know everything about the jewelry industry. During our conversations, she hinted at her desire to make rubies, too. She had the qualifications, but I still cautioned her and outlined all the roadblocks that stood in her way, including the built-in distaste people in the trade had of anything made by man. I think she took this as a way to keep her out of the business, and the relationship never turned into a friendship, but one of fierce competition and distrust.

When Osmer approached my cutters in Hong Kong to do her work, I objected, naturally. When she was able to buy supplies at the same price as me but at 10 percent of the volume, I complained and had it changed. When she dropped a prime booth in the Tucson Gem Show and I was first in line with seniority to take it, I did. It was just a good business decision, but she thought otherwise. It was her choice to leave; I did not push her out.

She put her business up for sale after that, with one stipulation: You cannot sell my business to Tom Chatham. The bitterness was and is, still there. It is too bad, because it was never personal from my point of view. I didn't mind the competition but thought that using my cutters and suppliers was a bit over the top. From what I hear, her process is still for sale.

Another, more positive, connection came

One of my joint ventures was with the Douros brothers in Athens, Greece. The facility was very primitive in every way – dirt floors and old wood benches – but they had a novel crystal growing process. I injected capital and made some suggestions, but they did not follow all my ideas.

about during one of the Tucson shows. Two gentlemen came up to the booth and handed me a large lab-grown ruby crystal. "Nice stone, " I said, "what's on your mind?"

"We have come for your help, we are the Douros brothers from Athens, Greece. I am John Douros and this is Angelo Douros," they told me. "We are growing flux rubies but find it impossible to sell! Can you help us?"

This is a common dilemma I have often encountered over my 55 years in the trade. People think making a product successfully is the key to success, however, it is only one piece of the puzzle. Yes, you need product, but more importantly, you need marketing and promotions to get your products seen. This was a common thread I saw from every scientist who came to us bearing gemstones they created. Experts in science, but amateurs in marketing.

Chatham had a huge advantage in this scenario. First, my father was the first to grow a commercially feasible emerald worldwide, and in the United States to boot. His accomplishments were written up in hundreds of articles, many pretty corny and poorly written, but the name was put out in the public eye for free. Then the marketers jumped on the opportunity, including Ed Coyne's Cultured Gemstones, Inc. New York, which later morphed into the FTC-agreed name of Created Gemstones, Inc.

Being second or third in any business is old news and hard to command attention, so they came to us for help. We turned down many because it just didn't make sense and we were not in the business of paying off competitors just to keep them out.

"Mr. Chatham, if you pay me $100,000, I will not introduce my emerald," they said, often holding a single low-quality crystal.

My father had a pat answer I use to this day: "Well sir, that's a very interesting proposal.

Left: The furnace was unique, based on a turning crucible to create convection while it slowly lowered to develop temperature gradients. **Right, top and bottom:** Tom and the Douros brothers inspect the month's production of rubies. This effort lasted about three years.

Bring me a kilo or two of good quality crystal and we can continue this conversation, otherwise, good luck!" They never came back.

But the Douros brothers were different. This was not a shakedown for cash, but a potential business proposal. If Chatham would help with capital, they would grow ruby for us. They had a promising product, but needed some technical help. I shared some critical information with them with the proviso that it was never to be sold to anyone. They could use it themselves; even compete with us, should we fail to use their crystals of acceptable quality.

I went to Athens many times and found myself in a beautiful land and inside the most ancient laboratory I have ever experienced. There were dirt floors, old wood benches, very primitive in every possible way, but they had a novel crystal growing process. I injected capital and made some suggestions, but they did not follow all of my ideas.

Scientists are stubborn, like my father could be, and follow their own paths, sometimes into oblivion. We worked with the Douros brothers for about three years, but the quality was not up to par in most cases, and the customers were demanding cleaner and cleaner stones without inclusions, so we parted ways, but on friendly terms. They both then retired. This experience, however, taught me a valuable lesson: Joint ventures can and do work, so I sought out other potential growers around the world to work with.

16
Navigating Possible Big Buyouts

Buyouts are always difficult and time-consuming, especially for a business such as Chatham. Buyers naturally want to know what they are acquiring, and you don't want to give away the secret process until it is paid for. My father was adamant about two elements of any deal: First, you must pay for my buildings and all equipment inside (lots of expensive platinum was involved). Second, he would not divulge the secret process upfront, only on payment.

Many companies tried to buyout Chatham Research Labs over the years, but we never initiated these discussions. My father was always open to a sales pitch for his business, but we had no idea of the huge business that was waiting to happen in crystal growth: computers, phones and thousands of other applications.

Union Carbide was first but failed because of antitrust law issues. We were all in agreement on terms and payout until one of Union Carbide's attorneys pointed out that they were buying out their only competitor, since they too had an emerald process as well as the Linde Star sapphire business. That killed that deal quick because they did not want to have a monopoly on creating emeralds and face the wrath of the FTC's anti-monopoly division.

Buyouts are always difficult and time consuming, especially for a business such as Chatham. Buyers naturally want to know what they are acquiring, and you don't want to give away the secret process until it is paid for. My father was adamant about two elements of any deal: First, you must pay for my buildings and all equipment inside (lots of expensive platinum was involved). Second, he would not divulge the secret process upfront, only on payment.

This made for some difficult times for my father – it was very distracting from his research and thus, annoying to him. Also, companies sought us out, we were not trying to sell. I started to step it up a few notches and volunteered to go to New York for meetings and take phone calls from attorneys concerning these interests in our business.

One thing the buyers would ask, which sounded very dumb at the time, but makes perfect sense now was: "Mr. Chatham, what else can you grow? What new crystals do you think are possible?"

My father's answer to these questions were pure Chatham: "Are you crazy? There are more than 100 elements on the Periodic Table. The permutations and combinations of possibilities are in the millions. You have to ask specific questions in chemistry, not foolish ballpark ones."

Keep in mind this is the dark ages of electronics, around 1971-72: pre-computer and solid-state electronics. No fancy calculators, no portable phones, no emails, just a landline phone or at best a telex machine, which we did not have.

"This is a great opportunity for my father, and I will do anything to see that it happens. You have my full support, but I do expect something out of this - a job. I know more about the mechanics of running this business than anyone, besides my father."

- Tom Chatham

What we didn't realize was that in the United States, the electronics industry and the Silicon Valley it would spawn, was about to give birth to a whole new world of computerization. At the center of it all would be special crystals grown by man, like silicon and germanium arsenide; weird combinations of elements that made no sense but that had special properties. The only way to tell if a particular combination had any special properties was to grow it, thus the question to my father, "What can you grow?" was not so stupid after all. But to my father it was like asking someone in 1571 if they thought air travel was possible. Stupid question!

Today, entire divisions of companies are devoted to this specialty, known as Materials Research. Stanford University has an outstanding group of Ph.D.'s that do nothing but grow different compounds to test. Miracles have been made, but it is a collective effort requiring deep pockets and vast resources. This was not Chatham Research Laboratories, but potential buyers felt Carroll Chatham had a mind for these problems and wanted him involved. We had no idea, nor did they share this future thinking with us. I'm not sure if even they knew what they wanted!

One deal in the late 1970s that came very close to being consummated – for nearly $17 million – was with W.R. Grace. I'm not sure what their main product was, but they did own a large jewelry manufacturer business too, so a fit was there. We opened the books

to them so they could scour tax returns and personal holding statements, but the process of emerald and ruby growth was to be kept a secret until it was paid for. We even devised an escrow account with a third party to hold the documents required.

Ed Coyne and his company, Created Gemstones, Inc. was written into the deal as a necessary aspect of the business. Someone had to do the marketing and advertising; there was no sense starting over.

The W.R. Grace attorneys also broached my involvement with the deal as I was driving two of them to the airport. I was an independent contractor of Chatham, not an employee, so they had no control over me, legally. "Tom, what do you think of this deal? What do you want?"

My answer was simple: "This is a great opportunity for my father, and I will do anything to see that it happens. You have my full support, but I do expect something out of this – a job. I know more about the mechanics of running this business than anyone, besides my father. I expect you to take care of me in some way...or you will have an instant competitor."

I smiled at them, ear to ear, and they got the message. "No problem, we want you too," they said.

During this courtship with W.R. Grace, I would travel to New York and meet with Ed and his salesman, Jerry Hulse, who had been with him for many years by then. Jerry previously sold Linde Star sapphire into the wholesale trade before Chatham Cultured Emeralds came around. They were widely successful, but pencil money to Union Carbide. Chatham Created Emerald was seen as a godsend to the dilemma of the declining interest in Linde Star. There was stiff competition coming out of Germany because Linde's patent rights and protections had run out. But, as I earlier stated, the buyout from Union Carbide was canceled, so they shut down the whole Linde Star division and dismissed Jerry.

The W.R. Grace deal was so close we could taste it. My father and I went to New York to cement the deal at the company's New York headquarters, which was in a huge monolith of granite that I remember to this day because of its unique shape: a sleek edifice with a curving front that soared at least 50 stories high. It looked impressive and very expensive. The $17 million probably wouldn't cover their electric bill for the year!

These were the first meetings my father had with any of the higher up executives who were reviewing the buyout. The executive we met with didn't like the deal and it soon felt like we were starting all over. He was acting very New York-ish, a tough guy who looked at us with contempt.

First off, he didn't want to pay up front. "This is bullshit. We don't pay for anything without complete disclosure," he said.

The W.R. Grace attorneys explained the nature of the deal to Mr. Big Shot, but he was not buying it. My father would not budge. More importantly though, was this declaration from the executive: "Should we make this deal and get over these little issues, we want you to move your entire operations to Buffalo, New York, along with your family, of course."

My father was astounded.

"Buffalo? Are you crazy? I am not moving any place, for any deal, and especially not to Buffalo," he declared.

The executive threw the thick file on the coffee table between us, declared the meeting over and advised my father to think long and hard about his decision. And with that the deal was dead, right there, with probably two years of negotiations down the drain. We got up and left.

At the elevator my father told one of the attorneys, an old friend, that Chatham was no longer for sale and to stop trying to peddle it. He was angry and felt very misled. I have never seen a deal go up in smoke quicker than this one, and it was W.R. Grace that initiated it! We had two attorneys from W.R. Grace trying to buy us out, until we told them it was rather counterproductive to bid against yourselves, no? So they persisted as one entity.

My father did agree to consult with W. R. Grace for a number of years, about five, but he would not relocate. He was not looking for a job and he definitely didn't want to become an employee of W.R. Grace. And Buffalo? My father was used to California sunshine, not eight feet of snow!

Following the debacle at W.R. Grace, we called on Ed Coyne, who was struggling since he bought out his partners (the Ipekdjians Bros.) and was counting on the deal going through to save his business. We let him know the deal with W.R. Grace was dead and said, "Let's get back to business selling emeralds and rubies!"

Shortly after this time period, probably in 1975, we started having a lot of trouble getting paid by Ed. Goods were not being shipped and invoices went unpaid. I assume a lot of words between my father and Ed were said out of my earshot because one night, around midnight, I get a call from Ed in New York, where it was around 3 a.m. He was an emotional wreck.

Liquidation Sale!

Chatham-Created Emeralds
$1,500,000.00
INVENTORY

UP TO 70% REDUCTIONS OFF REGULAR PRICES ON CHATHAM-CREATED EMERALDS AND CHATHAM-CREATED RUBIES IN ALL SHAPES, SIZES AND QUALITIES IN FACETTED STONES, CABOCHONS, CALIBRE AND MELEE SIZES, CRYSTAL GROUPS AND CUTTING ROUGH.

NO REASONABLE OFFER TURNED DOWN!
ACT NOW BEFORE SELECTION IS DEPLETED!

This is a bona fide liquidation sale due to the fact that Mr. Carroll F. Chatham, the producer, has discontinued his relationship with CREATED GEMSTONES INC. and has advised us that he intends to offer his product through other outlets.

5 DAY MEMORANDUM SELECTION AVAILABLE FOR A LIMITED TIME ONLY!

WRITE! WIRE! CALL!

CREATED GEMSTONES INC.
Formerly Exclusive Distributors of Chatham-Created Emeralds
574 Fifth Avenue
New York, N.Y. 10036
(212) 582-1790

Ed Coyne tried to run this advertisement in *JCK* magazine after we stopped doing business with him because he didn't pay his bills. My father threatened to sue if it appeared in print, so it never

"Tom, you have to get your father to cooperate and work with us," he said – or words to that effect. I told him he was calling the wrong person: "Call my father, he owns this company, not me." He never called my father, a big mistake in my opinion.

We had lots of family meetings about the situation with Ed and Created Gemstones Inc. in the days that followed. I was in favor of starting our own marketing corporation. The others agreed and we started Chatham Created Gems, Inc. with headquarters in downtown San Francisco. Chatham Research Laboratories was not a part of this new corporation, but a supplier.

J.P. Cahn, who wrote one of the best stories I have ever read on Carroll Chatham for True magazine, was hired as the manager. He hired Donald Middleton as a salesperson and advisors were brought in who told my father that money would have to be spent on advertising and making contacts if this was to succeed.

Ed was furious when he learned about our plans, to say the least. He threatened to run a going-out-of-business ad in *JCK* magazine, a leading trade book. "CHATHAM IS CLOSING ITS DOORS: 50% OFF ALL INVENTORY" if he was not going to be involved anymore. The publisher of the magazine called my father and asked if he approved of the ad.

"No, and if you do run it, I will sue you," was my father's response. "Chatham is not going broke or out of business, we just stopped doing business with Ed Coyne because he won't pay his bills!"

The ad never ran and was probably a bluff. Typical New York strong arm stuff.

We took legal action against Ed for nonpay-ment, but the only thing we got out of it was an offer to return rough and cut stones. I went to New York and sat down with Ed to examine the goods, which turned out to be all the lowest quality of every shipment we made. I refused the offer, and the case was never resolved. It just wasn't worth the cost.

I think it was a wise decision to just walk away. There was a substantial sum of money due, but you just must figure the costs involved in litigation to recoup it. Olive branches are hard to find in New York. Ed continued his advertising business, selling other created gemstones, but never really was able to duplicate the success of Chatham Created Emerald. At the time of this writing, he is in his 90s and still active. Despite our differences, we are friendly when we pass at trade shows.

17
The Difficulty of Being Independent

Chatham Created Gems, on the other hand, needed people who understood how to facet properly in calibrated sizes to fit manufactured mountings. This meant retaining an average of only 20 percent of the rough, a huge differential from natural stones, and made it possible to offer 5,000 stones of identical size and proportions to the big chain stores.

Chatham Created Gems, Inc. began trying to sell rough emeralds and rubies worldwide with a sales force of one: Don Middleton.

The crystal clusters in freeform style, like a druzy quartz, did pretty good and sold for $4 per carat in 1,000 carat lots. Many goldsmiths loved these free-form crystals, and we sold a lot of emerald and ruby in these uncut free form styles. Gold was still fixed at $35 per ounce, so big heavy rings and pendants were manufactured by a handful of buyers.

We were not so lucky selling the rough crystals to be cut into faceted stones. Don Middleton, went to many countries seeking buyers, but few people in the natural stone realm have the talent to "read" rough and know how to cut it into faceted stones and those who did know the business were not interested in our lab-created product.

Since Middleton was unable to develop many buyers for our rough crystals around the world, we decided to cut the rough ourselves, just like Ed Coyne had done. Few people in the U.S. trade at the time knew how to cut and polish gemstone rough – New York just didn't have the capacity or the talent to handle our growing production anymore. At first, we worked in Germany where many colored stone cutters still existed. We approached famous names like Petsch, Ruppenthal, Haubert and Golay Buchel of Switzerland, and all cut for us at one time or another, but they did not produce cuts to the standards we needed, which were different from natural gemstones.

The finished yield of weight of natural stones is all important, often at the expense of proportioning the finished stone for maximum beauty. On average, a natural stone cutter will retain 50 percent of the original weight of the rough and build a mounting around whatever size the stone ended up being. Things have changed somewhat today from the 1970s, but this is still the norm in places like Bangkok, Bogotá or Sri Lanka. Many stones must be re-cut because of this philosophy of saving weight.

Chatham Created Gems, on the other hand, needed people who understood how to facet properly in calibrated sizes to fit manufactured mountings. This meant retaining an average of only 20 percent of the rough, a huge differential from natural stones, and made it possible to offer 5,000 stones of identical size and proportions to the big chain stores. Even today, this is unheard of in natural stone cutting, diamonds included. It could be done but isn't, because these bad habits are too ingrained in even the best of faceters.

This is a sample of a by-product of growing large single crystals. They were sold "as is" to rock hounds and goldsmiths who made elaborate pieces of jewelry very popular in the 1970s before the price of gold shot up. Credit: Orasa Weldon.

I was spending half my time in the corporate offices of Chatham Created Gems, Inc. in downtown San Francisco and half in the two labs about four miles away. My brother John was taking over part of my responsibilities in the laboratories. I became pretty good at sorting and grading the cut stones into the standards we had established: Gem, Fine, A and B. Everything was cut to calibrated sizes to fit standard mountings.

I rejected almost all the stones on one lot that Golay Buchel cut. I thought they were poorly cut with open fissures, extra facets, lumps, etc., not at all like our agreed quality. I sent them a telegram to report my findings and request that all the stones be re-cut. They responded by saying they would be happy to recut them – at a 100 percent increase in cost per carat. Switzerland had just priced itself out of the market.

France was too expensive and no one in Italy cut stones. Middleton was running out of options, and I was running out of patience.

"Dad, I am getting fed up with Donald Middleton," I told my father. "He is pushy, acts like he is in charge and is not solving our problems in cutting. I want to be president of the corporation so I can tell him where I stand and what he should do, with authority."

And just like that my father made me president of the company at age 25. (I looked about 18!)

On his return from overseas, I informed Don of the changes. I could tell he wasn't too happy, but he didn't say anything. Since he was unsuccessful in finding any new cutters for us in Europe, I told him I was going to Asia, the developing area for cutting and jewelry manufacturing. I left three days later;

it was 1973.

We had already been approached by one Korean company with sales offices in Los Angeles. We sent rough overseas and got back thousands of stones, each individually wrapped in tissue paper that took two people three days to unwrap, so Korea was my first stop. The factory was set up in the town of Iri, about 100 miles outside of Seoul. This was my first visit to a cutting factory. It was dark and cold and officials from the Korean government were waiting for me.

"Mr. Chatham, you shipped in 5,000 carats of rough material, but the shipment out is only 1,000 carats. Are these people stealing from you?" they asked.

This factory was in a controlled import/export area that allowed tax-free status in a blocked off area surrounded by military guards, so it was watched carefully. Language was a problem, so I knew my explanation had to be very graphic. I pointed to the drain in the floor and said, "The missing material is down the drain, no one is stealing." Everyone was confused so I walked over to one of the cutting saws and scooped up a handful of sludge from the bottom.

"This is waste product; it is part of the process of cutting and must be considered when evaluating the business," I said. They accepted this explanation to the huge relief of the company president. If they thought he was stealing, they would lose their right to work in the duty-free zone.

I then inspected the operations and asked to see how the stones were handled for shipping. They showed me that when the cutter is finished with a stone, it was brought to a window where the cutter gets a mark by his name. Then the stone was handed to another person who weighed it and recorded

it in a book. Then it was handed to another person who measured the stone's dimensions and noted it in a book. Then the stone was handed to another person who wrapped it in tissue paper. Finally, the stone was handed to another person who put the wrapped stone in a box with others.

That is five steps – five people to handle one stone. But it kept people employed, a common practice in Asia and other parts of the world. I made a suggestion: Add one more person at the end who unwraps all these stones. They had no idea of how expensive it was for me to have someone unwrap the stones in the United States. They got the message and began packing the stones by size in plastic bags.

I soon stopped getting shipments from them and had to return to find out why. The factory was cutting all cubic zirconia (CZ), the brand-new diamond imitation and they were getting $3 per carat to cut it (it now sells for pennies per carat in cut stone). This was the same price I was paying them to cut our emerald, but the CZ was much easier to cut because it was flawless.

I told them how disappointed I was in their abandoning me without notice. I told them the CZ business would not last and not to call on me for work when it happened. Of course, the CZ business only lasted a year, and they came back, hat in hand, begging for work. I told them I would not support a company who did not support me, that they let me down once and might do it again. I refused their generous offer of half price for new cutting and they went out of business.

I needed to search elsewhere for cutters and went to Taipei, Hong Kong and Bangkok.

18
- - -
Doing Business in 1970s Hong Kong

The most expensive hotel on Hong Kong at the time was the Le Méridien Hotel … All I was interested in, however, was using their phone system, which was free. I couldn't believe it …. It was also convenient because I believed that most gem traders were on Hong Kong Island.

To visit Hong Kong in the 1970s was like a time travel experience.

On the one hand you had modern skyscrapers going up right and left, and on the other, rickshaws running on the roads. Every cliché you ever heard about China was true. Millions of people running around on a small island, all just trying to make a buck. Extreme wealth surrounded by abject poverty.

I had no idea where to start. I was staying at the Sheraton Hotel on Kowloon, part of the China mainland with the island of Hong Kong a quarter mile offshore. Transportation on Kowloon and Hong Kong Island was very cheap: taxis ran about 20 cents for the first drop of the meter, and you could go for miles on that. The Mass Transit Railway ran through an underwater tube connecting Kowloon to Hong Kong Island.

The Star Ferry was another great way to get to Hong Kong Island. First class seating on top was 20 cents and second class below was 10 cents. The government almost killed the Star Ferry system when mass transit came in – until somebody pointed out how much people loved it. Yes, it was slower and took 10 minutes to cross the bay, but it was peaceful and gave you time to reflect, compared to staring at the face in front of you as the train speeds under water at roughly 60 mph. Thankfully, the Star Ferry was saved, just like the San Francisco cable cars. Somebody had the good sense to see this was a great tourist attraction, not just a municipal service. I still enjoy using it to this day, whenever I am in Hong Kong.

Luckily for me, English was commonly spoken in Hong Kong, due to the lengthy British occupation from 1841-1997, except for when Japan occupied it from 1941-45 during World War II. It was strange to hear a Chinese person speak with a British accent, but this was very common.

The most expensive hotel on Hong Kong at the time was the Le Méridien Hotel with huge expansive lobbies and plenty of lounge chairs to have afternoon tea and crumpets. All I was interested in, however, was using their phone system, which was free. I couldn't believe it. (The American-style Sheraton, where I was staying, did not offer this service.) Le Méridien was very old school, but quite proper, so I needed to wear a coat and tie to use this lobby/office. It was also convenient because I believed that most gem traders were on Hong Kong Island.

I used the phone book and started calling anyone in the Yellow Pages who might have knowledge of cutters in the area. This was a time my

Hong Kong today, much like it was when I first visited in the 1970s, is still a dichotomy of mixed values – with the super-rich melding closely with the many not-so-fortunate people living in wooden boats their entire lives. Hong Kong has long been one of the largest (with a population of 10 million people) and most expensive cities in the world. Credit: Adobe stock.

well-known name paid off because most people accepted my calls. I found a few cutters and made appointments to visit them.

I also found an ad for Gilson Opals, made by Pierre Gilson, who had appointed an agent in Hong Kong to represent him, so I called them too, just to say hello. Mr. Lo answered, and I introduced myself, but he knew all about Chatham emeralds and invited me over to have lunch. He was a few years younger than me, so we were both just starting out. We discussed cutters and markets and he introduced me to my first dim sum meal. I had no idea what I was eating, but I think it was pulled pork inside a dumpling with a sweet sauce.

After lunch we said our goodbyes and he said something that struck me as very unusual for a competitor to say, "Nice to meet you, Mr. Chatham, I hope we do a lot of business in the future." I came away thinking it was just his use of English that was off slightly, but it really stuck in my head for all these years. I had no idea how meaningful his statement would be at the time. Eventually this man would run our entire Asian operations and cut millions of stones for us.

Next, I went to Bangkok, the world's center of ruby and sapphire cutting. I stayed at the Mandarian Oriental, the number one-rated hotel in the world for 20 years, on the Chao Phraya River. I had a suite for $45 per night and a request: "Please do not tip the wait staff, it is not our custom."

To say Bangkok was cheap is a major understatement. Also, because it is right on the equator, it is swelteringly hot in the summer months, March through October. The temperature is basically 93 Fahrenheit all year long, but the summer humidity makes it feels like 120 F. When the humidity drops in the winter months it feels cool – but it is still 93 Fahrenheit.

{98}

I visited a few cutting shops and stone dealers, some run by former American military who never left after the Vietnam war. I met Dick Brown, an ex-GI in the sapphire business, who had an office in the center of the red-light district, sort of strange bedfellows but it worked. By day the area was normal, with shops selling trinkets and assorted fruits and nuts, but after dark, the place was a den of iniquity, sex, drugs and floor shows I will not describe. Just about anything you have ever imagined went on... all night long. I did not stay up to watch. Dick was not very helpful but did become a good friend over the years and I had many beers at his place.

Overall, I found the cutting in Thailand to be too crude and more focused on saving stone weight, not bringing out the best of a stone. Many a stone has been ruined in Bangkok cutting shops to save a half carat and many, if not all, must be recut. Cutting stones to a certain size – known as calibration – was unheard of then, and it is only in this century that it has improved, but not entirely overcome. I found out that teaching an old dog new tricks was near impossible in Thailand, so I left and returned to Hong Kong.

I found one cutting shop run by Julius Chang at Superior Jade Factory in Hong Kong. They did not have any jade to cut, so we made a deal with them to cut our stones. We started out slowly and soon had a humming factory producing beautiful stones. Every day we would get small parcels of cut stones, exactly to our specifications. I believe this lasted about one year, then abruptly stopped.

International telephone service was very poor at the time and telegrams were difficult because of language problems. Some words just don't translate as you wish, and twisted meanings could come out of them. So, I got

back on a plane to Hong Kong – in coach – taking about 15 hours of airtime. If you hit a head wind, you had to stop in Manila or Taiwan to refuel, adding another two to three hours. I made that trip hundreds of times and racked up over two million air miles on United Airlines, mostly in coach, unless I got bumped up to business class. In the beginning, that was common, later, as more and more people flew to Asia it became non-existent. I met some people who flew to Hong Kong every week or two. They earned that upgrade; it is a grueling ordeal, and you lose half your life in the air.

When I arrived at the offices of Superior Jade Factory it was empty, no employees. Julius told me that "The workers demanded a 50 cent (HK) an hour raise so I fired them all!"

I was astonished and said "Oh, that's smart, now you have no workers and I have no stones. You should have called me, and we could have worked out something, but this is not going to work. You cut off your nose to spite your face!"

He had no idea what I was talking about, but felt he had to stand up to his workers. A 50-cent raise was considerable back in 1974, when the average wage was about $4 per hour in Hong Kong currency, or less than $1 U.S. dollar. I would have worked with him on his fees to keep stones coming in and dealt with the workers in some way to appease them, but never had the chance. We parted friends and I also helped him and his family emigrate to the United States. It was a long process, about seven years, but successful. They were very grateful for my assistance.

So out I went again to find cutters.

19

Oswald Dallas and His Cutting Factory

"Well, Mr. Dallas, knowing what gemstone mining can be like, do you have a steady supply of rough to cut, or are some of these men sometimes sitting idle?" I asked, pointing behind me at a huge plate glass window, where about 50 workers hovered over faceting machines. "What I am asking you to consider is using those idle workers to cut our stones and maximize your profit."

– Tom Chatham

Going back to my temporary Hong Kong office in the Le Méridien Hotel lobby, I searched the Yellow Pages again for any leads. I hit on gemstone dealers and manufacturers and found an important sounding company called Dallas & Co., a subsidiary of Jardine Matheson. I looked up Jardine Matheson and found out that it was a huge conglomerate of companies based in Hong Kong, dating back to 1832. I figured this was like going to General Electric to buy a light bulb, but what the hell, I had nothing to lose.

Like many offices I visited, it was inside a building that looked industrial on the outside but had nice modern offices inside. If a company had no reason to be in the downtown section of Hong Kong, where rents were high, an industrial building was adequate. This is still the rule today in Hong Kong.

I was led into a formal office with a big board room table in the middle. At one end sat Oswald Dallas. He did not get up to greet me, but pointed to a chair and said, "Take a seat, want any tea or water?" Without waiting for an answer, he said he was seeing me because he knew the name Chatham Created Emerald and was giving me a courtesy audience. He was a real charmer and obviously of British nationality.

Mr. Dallas, as everyone in the company called him, was about 5'4" and a little overweight. He wore a black suit, white shirt and tie, complete with French cuffs and cuff links. In 30 years, I never saw him in anything different. He was very British and acted like it. Everyone was beneath him, especially the help.

"Well, Mr. Chatham, why are you here?" he growled. "I am a busy man."

"Thank you for seeing me, Mr. Dallas," I said. "I am in Hong Kong looking for cutters to facet our emeralds and I heard you are the best in Hong Kong." I thought this bit of BS would soften him up.

"Why in the world would I want to cut synthetic emeralds?" he spat out. "I cut natural emeralds for our retail stores here in Hong Kong, owned by one of the oldest companies in the world, Jardine Matheson, of which I am a director."

I pulled out a few large crystals of our emerald and put them in front of him. "I need your expertise at cutting natural emeralds because our emeralds need that practiced eye to yield the best cut stones. As you can see, our rough has many flaws, like a natural stone and it cannot be

{101}

"We can't cut fast enough"

Oswald Dallas is generally acknowledged as the man who established the business of cutting faceted gemstones in Hong Kong.

He was born in Shanghai, China. His mother was a gemmologist and his father had a trading company, Dallas & Company, which was started in 1853 in Shanghai by Oswald's great uncle, Alexander Grant Dallas.

The Dallas family left Beijing and moved to Hong Kong in 1948 and continued with the family business. In 1954 the family moved to Africa to set up a textile factory. It was during this time that Dallas Gems and Minerals Co Ltd was established in Rhodesia, now Zimbabwe.

Oswald Dallas returned to Hong Kong in 1964 to reorganise a trading company and also to establish Dallas Gems and Minerals Co Ltd in Hong Kong which was supplied with rough gemstones from the company in Africa.

In 1974 it was agreed to sell 51 percent of Dallas Gems to Jardine Industries. The balance of the company was sold to Jardine Industries in 1980, when Oswald Dallas retired.

Jardine Matheson & Co Ltd, the parent company of Jardine Industries, withdrew from the jewellery business in 1983, and asked Oswald Dallas to temporarily assist Jardine Industries in closing Dallas Gems.

Stock had to be disposed of and people had to be found jobs. So he decided to buy some of the stock, reemploy his old staff and reactivate Dallas & Company. "I've had enough of retirement: I need an office to go to," he told friends at the time.

Three years after having retired, he found himself back running his own gemstone cutting factory.

Jewellery News Asia talked to Mr Dallas at his factory in Kowloon.

Oswald Dallas: "we don't compete"

What is your production capacity?
Mr Dallas: We can produce about 2,500 carats of cut gemstones a month.

What sizes do you cut?
Mr Dallas: There is a time factor involved. We don't like to cut two millimetres or three millimetres, because to cut such small stones takes as long as a large stone.

How has the trend towards lower-priced jewellery containing small, low quality stones affected the business of your company, which specialises in precision cutting and larger sizes?
Mr Dallas: We have to be very careful. We pay our cutters the highest rates in Hong Kong, and also provide medical benefits and free lunches. We cannot cut cheap stones. Take, for example, a garnet, which is a cheap stone; if we were to cut garnets we would lose money.

Our orders have been increasing all this year. We are employing more cutters and may have to move to larger premises.

With higher costs, how do you compete with other companies?
Mr Dallas: We don't compete. We are quality, precision cutters. Accordingly, we have our prices and we maintain that our cutting is the best.

What is the difference between cutting man-made gemstones and natural gemstones?
Mr Dallas: With natural gemstones, the size of the rough dictates the end result. The stone is cut to get the best yield, so you can't cut a stone to meet an order.

But with man-made stones the situation is the reverse: the shape, quality and size are all decided by the jewellery manufacturer and the stone is tailor-made to meet the order. We cut an enormous range of sizes and shapes, no two orders are ever the same.

Which is in greater demand — natural or man-made stones?
Mr Dallas: Almost 90 percent of what we cut is rough from Chatham Created Gems Incorporated in the United States. Demand is so strong we can't cut the rough fast enough.

At a time when the jewellery market remains depressed, why are Chatham stones selling so well?
Mr Dallas: That is a question I have asked Thomas Chatham. He says he doesn't know the answer, but his business keeps increasing year after year.

My view is that people do not wear expensive jewellery

Preforming stones

any more. You can tell that from the way they dress. Chatham stones are not cheap, but they are less expensive than natural gemstones, so they meet a demand in jewellery that exists today.

Why does Chatham, which has its factory in the United

The cutting section of Dallas & Company's factory in Kowloon

Jewellery News Asia — September 1985

The big three in gemstone demand and value: blue sapphire, emerald and ruby, photographed in the mid 1980s Chatham tried to produce only what was popular and expensive in natural stone.

handled any differently."

"We don't cut synthetic stones, why would we?" he said again. "We mine our own rough in Africa and we cut it."

"Well, Mr. Dallas, knowing what gemstone mining can be like, do you have a steady supply of rough to cut, or are some of these men sometimes sitting idle?" I asked, pointing behind me at a huge plate glass window, where about 50 workers hovered over faceting machines. "What I am asking you to consider is using those idle workers to cut our stones and maximize your profit."

Bingo. I had hit the magic button.

"Perhaps you have a point, Mr. Chatham, we do have down time and I must pay these people regardless," he said. "Send me a few hundred carats and we will decide then."

And so began a 40-year relationship with Mr. Dallas. We soon overwhelmed his capacity and as his natural emerald cutting dwindled, the Chatham business was exploding. It was an opportune time for Jardine's as well, as its jewelry business was faltering and on the brink of bankruptcy. The jewelry business has always been very competitive in Hong Kong and rents were becoming unbearable, while mainland Chinese companies began opening many jewelry stores with cheaper labor.

People from all over the world were discovering Hong Kong's quaintness and with jet travel more common, it was a very reachable destination, especially for the evolving wealthy Chinese from the mainland, just an hour's drive away.

Crossing into mainland China as a foreigner requires a visa and at the time, only a walking bridge existed over the border. There used to be two sections of Hong Kong: HK island and the "new territories" or Kowloon. This was all under British rule. To go into mainland China was a big deal as it was a totally different government with a huge river separating the two. People from HK were not allowed. Papers were inspected and baggage was searched. Even local mainland Chinese had to have a special permit to go in.

There are three sections of Hong Kong: the island itself, the "new" territories, and the Kowloon Peninsula, which were all leased by the British and under their jurisdiction. Crossing into mainland China as a foreigner was a big deal, because it was a totally different government than Hong Kong and required a visa. Papers were carefully inspected, and baggage was searched so it was not easy to do business there until years later when the Chinese government opened up the city of Shenzhen to foreign business entities.

Mr. Dallas was a unique individual. On the outside he was British, but on the inside, he was Chinese. He was born in China, spoke Chinese very well and was a tough negotiator because of that. The British aren't known for being tough negotiators, but the Chinese have it down pat. So when you were dealing with Mr. Dallas, you were dealing with a very skilled Chinese merchant.

Born in 1904, Mr. Dallas had a very interesting life. His family was quite well off and raised thoroughbred racehorses in China. Dallas, being of slight stature, was an accomplished jockey in his younger days. The family fled to Rhodesia during the revolution of 1949, leaving everything behind, including a big mansion, stables and valuable horses. They were lucky to get out with their lives. This was the beginning of communist rule when the Chinese leader, Chang Kai-Shek was overthrown, and the Peoples Republic of China (PRC) was created. Chang Kai-Shek and his followers fled to Taiwan and created a new government, much to the chagrin of the PRC who still consider Taiwan as part of China.

Mr. Dallas got his education in mining and cutting of gemstones in Africa, which was just becoming a major source in the diamond and gemstone world. Because of this experience, he caught the eye of Jardine Matheson executives, sourcing materials to sell in their Hong Kong jewelry stores.

Years later, after China settled down, Mr. Dallas returned to his homestead in China.

It was a sad homecoming. The house was in shambles – many rooms had been turned into little apartments and the grounds left to fallow. Missing his Chinese homeland, Hong Kong was the next obvious choice for him to live, permanently.

Over the next few years, Mr. Dallas and I became very good friends. I was the only one who could call him Oswald. Not even his lifelong friend, Mr. Wong, called him Oswald. I, of course, was just Tom.

After many years of cutting millions of stones, Mr. Dallas and Jardine Matheson, his parent company, asked if I would take over the ownership of the cutting factory – at no cost to me. Jardine Matheson Inc. wanted out of the jewelry business. All they asked was for me to just keep the employees and it was mine. The labor laws in Hong Kong are quite specific: Fire someone or close a business, and you owe each of the employees affected one month's salary for each year worked. Some employees had been with Mr. Dallas for more than 40 years. Mr. Dallas agreed to stay on and oversee the cutters.

This continued for at least another five years until Hong Kong priced itself out of the market. McDonald's was offering more than we could afford for new employees, so we decided to partner up with another company and move to Shenzhen, China, then a town of only 5,000 people about an hour's drive from Hong Kong.

I could have ducked out of that obligation to pay each employee one month's salary for each year worked when we eventually did close it down in Hong Kong, but I didn't. It was a substantial sum of money. None of the cutters were allowed to work in mainland China.

20
Meanwhile, Back in the USA

One of the strange rules we had in the labs, which always made Ph.D.'s upset when they would come around looking for work, was: "Sorry, we don't hire smart people like you – it's too dangerous." Keeping a secret process secret was not easy and we had certain rules for outsiders: no one with a degree was allowed to work for us, especially those with a degree in chemistry.

Business in the U.S. was thriving.

We had started a relationship with a company called Heller Hope in New York, the originator of the Hope Star and the nemesis of Linde Stars made by Union Carbide, to sell Chatham gemstones. They hired Jerry Hulse from Ed Coyne at my suggestion and we split the U.S. down the middle, between Heller Hope and Chatham.

Jerry and I hit it off well, going out to dinner or drinks and talking shop every time I visited New York. He was about 20 years my senior and an ex-air force pilot who had flown many missions in World War II flying a B17 Flying Fortress bomber. He was shot down twice with hardly a scratch – a miracle at the time, as few pilots lived beyond 20 missions. He never married, to my knowledge, and I never asked. He loved to drink and knew everyone in the jewelry business, from secretaries to presidents. Everyone loved Jerry. These talks and meetings with him gave me an inside view of Created Gemstones Inc. and a better perspective of the jewelry business overall.

Jerry and I did many hundreds of trade shows together, constantly pushing the retailers to accept our created gemstones, and we became lifelong friends. He knew everybody in the industry and that opened a lot of doors for our emerald business. It was not an easy sell, but little by little we grew, and with office staff in San Francisco and New York answering calls for loose cut stone, we reached a point of more than 200,000 stones sold in a year.

That's a lot of cutting and Dallas in Hong Kong couldn't believe the demand. We even tried expanding into Bangkok for cutting, but that never really worked out. They just didn't have the quality control we needed – and refused to learn it.

It is very hard to break old habits in faceting. To a cutter in Bangkok, yield was everything. There is always room for negotiating in the natural stone business, so if a stone was lumpy on the bottom, the weight loss on re-cutting was factored in. This is not how we looked at cutting a Chatham stone. I could not sell poorly cut stones, so every poorly cut stone meant that we would have a 100 percent loss on that stone. We learned it was better to take an untrained person in Hong Kong and then China, and teach them proper faceting and standards we needed, rather than try to have them unlearn bad habits.

Family members started joining the company to help handle the

A Chatham lab-grown sapphire advertisement. At this time, we reached a point of more than 200,000 stones sold in a year.

increasing load. My brother John was already online, and he hired his son Keith and my son Rod, along with others to help out in the labs. We also drafted my daughters, Serena and Christa to handle the ever-increasing phone orders from retail stores coming into the corporate headquarters in downtown San Francisco. My wife Dianna joined me in the 1980s after a very successful run with Pacific Bell and handled all Human Relations and general oversight of the employees. This number got up to around 25 people when all sales offices were considered, just in the United States. Overseas we employed hundreds of people.

I realized that hiring family is not always a good idea. Some made it, some didn't. My brother fired my son Rod because he could never be found, and Christa quit to chase some Air Force guy across the United States. Everybody ended up happy, but it does not work to draft people to work just because you know them and trust them. They must want the job, not just want a job. My daughter, Serena has stuck it out and is still with us after more than 30 years. Two out of five – not bad.

One of the strange rules we had in the labs, which always made Ph.D.'s upset when they would come around looking for work, was: "Sorry, we don't hire smart people like you – it's too dangerous." Keeping a secret process secret was not easy and we had certain rules for outsiders: no one with a degree was allowed to work for us, especially those with a degree in chemistry. Jobs were compartmentalized so no one would see the entire operation. None of the chemicals were labeled and all the instructions were coded: "Put 1230 grams of bottle A in unit 15. Put 230 grams of bottle B in unit 12."

My father never had an employee – it was all family – and this was one of his reasons; people have a price. When I took over, we had no choice but to hire outside people, but they were never well educated.

We never had a leak of information.

My wife immediately took the roving GIA courses, this time taught by Harry Stubbert and Peter Hess. Again, the Chatham name made these teachers nervous. At the completion of the two-week classes, we took Harry and Peter to lunch at the Cliff House restaurant on the beach overlooking the Pacific Ocean.

These two were about as green as they get and were impressed that I would take them to lunch. It was the beginning of a 30-year friendship that ended up with Stubbert taking over Chatham Created Gems years later. Harry soon rose through the ranks of GIA and became CEO of the instrument division and ran it very profitably. Every student needed a microscope and other identification instruments and who better to supply them than GIA?

Unfortunately, Peter Hess died from a tragic fall off a ladder at a very young age.

Harry remained with GIA for some time, and we often got together at trade shows when both of our companies exhibited in them, including the Hong Kong show where we exhibited for many years.

As was usual with any trip to Hong Kong, a formal dinner with Oswald Dallas and the head cutter was required and always consisted of top-end Chinese cuisine consumed over several hours. Oswald invited some of his old Chinese pals, one who was imprisoned in China for 30 years for political reasons in 1949, and P.W. Chew, who had a chauffeured Rolls Royce drive him and his wife around. I never knew what he did to live so lavishly, but he was obviously very successful at whatever it was.

I invited Harry to one of these dinners at the Regent hotel in Kowloon, my home base in Hong Kong for many years. I watched it being built from my room at the Sheraton and it blocked my view, so I moved. It was right on the Hong Kong harbor and looked out to the Hong Kong Island. Very old Chinese junks would be constantly sailing back and forth with their cargos of sand or fruits. It was very colorful and a joy to watch every morning. All that went away after 1997, when China took back Hong Kong and Kowloon, breaking the 100-year lease they had previously agreed to with Great Britain. People in Hong Kong today are not happy with the communistic rule.

At dinner, I secretly told Oswald to give Harry the head when the Peking Duck was served. Harry was way out of his element at this table, being 12 years my junior and I was still quite young, maybe 35. I had been well exposed to this type of dinner in Hong Kong before: all very formal and stuffy around a huge round table with a lazy susan in the middle. When the head was placed ceremoniously on Harry's plate, one of the other guests told off the waiter in Chinese. The waiter explained, in Chinese, that Oswald had ordered it done and said it was a joke suggested by me. P.W. Chew then told Harry the head had the best parts and explained how to eat the tongue, then the cheeks.

So, Harry ate the head after all. He could have killed me.

21

African Adventures

I got a funny feeling and slowly turned around to find a group of adult baboons lined up behind us – we are between them and their young. Baboons can easily kill you. This is what caught the attention of the running meerkats ... they could sense trouble. They all stood on their back legs because they are only about 15 inches in length, just staring at the adult baboons.

Harry Stubbert and his wife Patti became good friends of ours over the years – we even took vacations together. One notable trip was an African safari in 1990 put together with our friend, Abe Suleman, who we both knew from the States where he had a loose gemstone business called Tuckman International. Abe was of Indian nationality but raised in Tanzania, Africa, and fluently spoke Swahili, as well as English.

He proposed a trip from Nairobi in Kenya to his town of Arusha in Tanzania. This sounded easy enough and even though it wasn't going to be a safari like Abercrombie and Kent would put on, he promised it would be an exciting adventure. Boy, was that an understatement.

Harry's wife was a travel agent at the time, and got us around-the-world airline tickets in first class for about $1,500 each. It didn't matter how many stops you made, just so long as you kept moving forward.

Just getting to Nairobi was a challenge: 12 hours to London, then another 12 to Nairobi. Nairobi was a slum city and dangerous, but Abe had family there, so we were safe inside a compound with high walls studded with broken glass.

The next day, Abe rented a Volkswagen bus and hired a driver. We were told to bring little because of a lack of space, so we all packed light. We had no idea what to expect but were told we had to do about 100 miles a day to be on schedule. I thought to myself, big deal, 100 miles is nothing – on a paved road. There were no paved roads in Kenya, just dirt paths and millions of animals. We felt like we were in a *National Geographic* special.

The animals didn't mind sleeping in the middle of the road, and we soon found out that lions don't move when you honk at them. Nor do elephants or hyenas. There were signs posted all over Kenya that said, "Do not leave your vehicle" and "Do not take a walk outside." Later, we found out why.

You have no idea how empty Africa is until you travel its dirt roads. You can drive for hundreds of miles and not see a soul. We always carried extra fuel, but food was tough to find. We passed occasional shacks selling things, but what they had was not very edible looking – some kind of unknown dried meat and nuts. Getting sick was not on our list of to-dos, so we kept going.

We stopped at some beautiful lodges on the Serengeti, but most had

An African sunrise looking out over savanna territory in Kenya. Credit: Adobe stock photo.

no food, water or gasoline. We had brought nothing, not even a salami or a bottle of wine, but Abe managed to scrounge up food, like chickens or beef, from local villagers. Luckily, we could buy wine and liquor easily. The lodges were more than happy to prepare whatever Abe found for us to eat because they had so little to offer us during our stay. One night we had wildebeest and it tasted like shoe leather. I don't advise it.

One odd sight was seeing Maasai warriors walking across the Serengeti 400 miles from anything with nothing but a spear for protection. We were warned, "DO NOT take their picture," just say "Jambo," which means hello in Swahili. Many were more than 6 feet tall, in bare feet and very colorful in bright red dress. They lived on the land and only drank cow's blood mixed with milk. We didn't join them. With permission and a few coins of the local currency we took some pictures. We encountered many Maasai people over our three-week journey.

Driving in Africa was exhausting. These were not nice flat dirt roads, but animal walkways full of potholes, washouts and boulders. The only one who knew where we were going was Abe, so we all had to just bear up until we saw some civilization or an outpost. It usually took us about 10 hours to drive 100 miles.

We also had a lot of mud and car problems. One day, trying to do our 100 miles, the engine just sputtered and died. I have had a lot of experience with VW's and they will usually run in any condition, so we had to figure out what was wrong. We checked the gas tank and battery, and each was OK. The engine cranked but there was no pop. So, I got under the bus and saw that the clear plastic fuel filter was full of mud. I think we had a hammer, a screwdriver and a pair of pliers. I

> "That [leopard] is one of the few animals that will eat you with pleasure, they kill people all the time. Get in your vehicle and stay in your vehicle and I will watch the leopard until you're out of sight."
>
> *- African park ranger*

took off the fuel line towards the engine and nothing came out. Then I took off the whole filter and gas poured out of the fuel line before the filter. We had a spare filter – I don't know why, maybe this was a common problem with local gas. With the new filter in place, the car ran like a top. I was the only one with any mechanical experience and without it we could have been stranded for days, if not weeks.

Next, the inexperienced driver hit a huge mud hole and got us bogged down in mud. We all got out to push the bus and while we were pushing from the back, the driver was racing the rear wheels in reverse, and we were covered in mud. We got him in the right gear, and we were able to get the car out.

Just then, a game ranger pulls up and walks over carrying a rifle. "What are you doing out of your vehicle?" he asked. We told him how we were stuck in the mud hole and just got ourselves out.

He pointed to a tree about 100 feet away where a leopard was 25 feet up on a branch just watching us, big paws hanging down – just like on TV.

"That is one of the few animals that will eat you with pleasure, they kill people all the time," he said. "Get in your vehicle and stay in your vehicle and I will watch the leopard until you're out of sight."

Do not ever go on a safari this way, trust me. One place had mud running out of the shower. I let it run for 20 minutes and there was still mud, but it was cleaner mud than the

mud I was wearing, so I took a mud shower. One morning in that hotel/motel we woke up with a full-grown, big-bull elephant looking in our window! I did not try to feed him. Kenya is not a petting zoo.

We stopped by some of the gem deposits on our way through Tanzania. One belonged to Campbell Bridges, the first person to discover Tsavorite (green garnet) at his Scorpion mine. Bridges was not home, but the mine was working. I looked around the land they had decided to dig on and could see for at least 50 miles in each direction. The open pit was about 100 feet across and 30 feet deep – all dug by hand because human labor was much cheaper than diesel. Years later, Bridges was murdered on this land by men trying to steal his mineral deposit. They were eventually caught and convicted and his son, Bruce Bridges, continues to run the mine today.

Onward we went, stopping at the occasional shack for a beer. Some had gemstones to sell, but nothing worth buying. We soon came to a low river and decided it would be wiser to walk it to make sure there were no deep holes and have the VW follow us. Patti, dressed in all white, was not walking in the river, so Harry gave her a piggyback ride over. Patti didn't quite bring the right clothing for an African safari, at least not the kind we were on. It was muddy and dusty with no place to clean up and by the time we got to Arusha we were all filthy, fed up with the trip and relieved to stay at a real hotel.

I got in the shower with all my clothes on, including my shoes, and turned on the water. Layer by layer I took my clothes off, down to bare skin. I got most of the mud off the clothes, but there was so much, it clogged up the tub!

We stayed in Arusha for a few days to contemplate our return trip. Nerves were a little

Elephants crossed our path on the plains of the Serengeti. You are advised to stay in your vehicle in Kenya or Tanzania. Animals are everywhere and some may attack. If a lion is sleeping on the road, drive around him and don't honk your horn!

frayed by this time, after travelling over 800 miles over bumpy, dirty roads, but we saw a lot of animals and only got in a little trouble a few times. We decided to go back to Nairobi via a different dirt road that took us through some new territory.

We stopped at a quaint one-story hotel that was raised off the ground for protection from animals. The beautiful polished-wood dining room was all open air with a gentle breeze coming through – luckily, we were not there in the hot season of summer, more like early spring. The big round table was set with a white tablecloth, fancy plates, glasses and silverware. We were the only people in the hotel dining room and felt like we were in heaven.

Our well-dressed waiter, who spoke English, put a large menu with all sorts of delicacies outlined, in front of each of us. Harry started first; "I will have a steak with fries," he said.

"So sorry sir, we are out of the steak."

"OK," said Harry, "make it a hamburger."

"So sorry sir, we are out of the hamburger."

"OK, why don't you tell us what you do have?"

"We have corn soup and bread pudding," the waiter said.

Why we went through all the song and dance was beyond us, but after what we had experienced in the last week, no one was surprised. "OK, we will all have corn soup," Harry said.

The corn soup was served, but before we could begin, a large worm dropped from the rafters into Patti's soup. She screamed, the waiter ran over and she pointed at the worm squirming in the bowl. The waiter calmy took

{113}

Tom and Dianna Chatham at the foothills of Mount Kilimanjaro with a rare sight in Africa behind them: a clear shot of Mount Kilimanjaro minus the clouds that often hover at its peak. We did not climb to the top.

a spoon, dished out the worm and walked away, not replacing her soup. Patti did not eat the corn soup.

The waiter cleared the table and began to serve the bread pudding. Out of nowhere, an adult baboon flies through the open window and attacks the table of bread pudding, jamming hands-full into his mouth as he runs around the dining room screeching and jumping on furniture, swinging from the rafters and having a good old time. The waiter is desperately trying to shoo him out with a broom and is lucky the baboon didn't turn on him as they are very dangerous. Finally, he gets the baboon out and the place is a huge mess. So, no bread pudding either!

Harry and I decide to go outside so he can have a smoke. We wander about 100 feet from the hotel and spot a tree full of baby baboons playing in the trees. We approached slowly hoping to get a good picture. Out of nowhere we hear this rumble and a huge mob of maybe 200 meerkat's (a type of mongoose but skinnier) run toward us, screeching to a stop and standing on their hind legs to look at us, or so we thought. They were very curious, but not coming closer.

I got a funny feeling and slowly turned around to find a group of adult baboons lined up behind us — we are between them and their young. Baboons can easily kill you. This is what caught the attention of the running meerkats … they could sense trouble. They all stood on their back legs because they are only about 15 inches in length, just staring at the adult baboons. I tell Harry not to run, "Just slowly walk away from the babies in the tree and don't do anything, anything, to scare them."

{114}

We carefully backed away while the baboons and meerkats all intently watched us, especially the adult baboons. Once we were clear – out of range and got back on the hotel porch – the mob of meerkats went on their way and the baby baboons continued to horse around in the tree. We did not venture out again!

The rest of the trip back was uneventful – until we came upon a swollen river about four feet deep and 30 feet across. We can't get around it and back track because we are short on fuel, so our only hope was to wait for the raging river to subside. There were about 50 vehicles on each side doing the same. All we had was a bottle of Sapphire Gin, so we drank it and watched. Abe figured it would be about five hours.

Some trucks got impatient and thought their height and weight would enable them to forge the river. Those who tried immediately began to be swept away, but luckily were able to back out. A school bus full of workers tried and nearly flipped over and got stuck. Four feet of water moving at a good clip is dangerous and powerful.

We waited all afternoon until the river was passable and the gin was gone. One by one, the bigger trucks made it through, pulling some of the cars as a safety precaution. The school bus was able to drive out and we finally were able to cross too. Thankfully, no one waiting was hurt.

Out next challenge was in the middle of nowhere at the border between Kenya and Tanzania. The little check point consisted of a tent with guards holding machine guns.

"Let me handle this, it may cost us," Abe said as he gets out and goes into the tent.

When he returns, he tells us that he bribed the official so he would not inspect our luggage and possibly steal something of value, but that we all needed to go in the tent with our passports. Inside, an overweight, bored guard was sitting behind a desk. He looks at each passport and verifies the person, then literally throws the passports at each of us. I was insulted and gave him a look he didn't like. The guard barks out something in Swahili and Abe answers and tells us to get out of the tent before he changes his mind and has all the luggage opened and rifled through.

Welcome to Kenya, the sign says. The bribe was $20. On to Nairobi with a brief stop in Mombasa.

From Nairobi we took a flight to Bombay – now Mumbai – where we saw miles of sheet metal and cardboard slum huts as we got closer to the tarmac; some were right up to the landing area. We almost missed our connection to Hong Kong, but thankfully made it just in time. We did not want to stay in Mumbai from what we could see of it from the plane.

Hong Kong felt like heaven on earth: Clean clothes, food, wine, clean beds. We were never so happy to be out of Africa. We have a lot of good memories of the trip but would never do it like that again. I recommend that you go with the professional guides and fly into the hotels; they all have little airstrips just for that reason.

I have visited just about every continent on the planet – racking up more than 3 million miles in the process – and visited many exotic places, but my most memorable has always been that African trip…after it was over!

{115}

22

Dealing with the Cartel in Colombia

We were each given a turn at the jackhammer, but when my turn came, I slipped a large Chatham emerald crystal in my hand. I pulled the jackhammer back and held the crystal up. I thought the head miner was going to have a heart attack, until someone quickly told him it was a joke, not a natural emerald I had dug out. That was the last time I pulled out any crystals.

▌ **Probably my scariest trip abroad** was to Bogotá, Colombia and its emerald mines.

▌ Colombia has long been the source of the finest emeralds ever found in history. Cleopatra and the Indian Maharajas were known for wearing Colombian emeralds; even Nero, Emperor of Rome, had a pair of eyeglasses made from Colombian emeralds to better see the gladiators fight to the death.

Natives have been mining Colombian emeralds for centuries. The mines are hundreds of miles from civilization and are in extremely difficult terrain to traverse. The mountains – thousands of feet high – are everywhere and all look like the same dense jungle. There is one dirt road that extends for hundreds of miles. After Spanish conquistadores invaded, pillaged and ran amok for many years, the mines were closed off and hidden in this dense terrain.

The mines re-opened sometime in the last century. Unfortunately, Colombia is also a major source of illicit drugs, such as cocaine and marijuana, which are either produced in Colombia or pass through on their way to the United States and other continents. The three principal mines, Muzo, Chivor and Coscuez, are run by warlords whose word is final. The Colombian government tried to run the mines at one point but could not control the constant thievery and unrest. In came the warlords, who bought concessions from the government and shared the profits to some degree. These warlords have the manpower, heavily armed with automatic weapons, to back them up. Because the mining areas are so large, it is difficult to guard everywhere, all the time. Thousands of local peasants, who are also armed, work the tailings of these mines on the lookout for the missed emerald or two. Fights and gunfire are common.

When I was there in the 1990s there was one rule to follow: If you find an emerald of any value, the mine owner has right-of-first-refusal. If he wants what you have found, he will negotiate, but the mine owner always comes out on top. If the owner doesn't want the stone, it is yours to keep. If the worker is caught with an emerald that he didn't offer to the owner first, he could be shot for stealing. There is no law and order in Colombia's jungles – only the smartest survive.

Few Colombians ventured out of the country with a sack of emeralds to sell because of the heavy taxes required to bring them back into the

{117}

Top: The entrance to the Muzo mine shaft, looking down into the mine. We traveled more than a quarter of a mile down the shaft in an open-sided lift. **Bottom:** The temperature and humidity rose by 50% by the time we hit bottom. Steamy, hot water seeps from the walls of the mine and soon everyone turned black and sweaty. The tunnels, dug over the last 200 years, stretched for miles and we were advised not to wander off. The public is never invited to these areas because it is very dangerous.

Muzo mine ore is constantly brought to the surface and immediately searched for emeralds. Everywhere you looked on the ground were thousands of little emerald crystals, too small for cutting. If you wanted one, you could just picked it up.

country if they can't sell them. So, stones were sold in Bogotá to foreign dealers who then marketed them in their own countries. The locals were getting screwed, price-wise. The government tried to change that by sponsoring an international symposium on the emerald business in Colombia in 1998.

One would think that after 70-plus years of growing emeralds in San Francisco, the name Chatham would be known in Colombia, but we never shipped into South America because imports of this type were forbidden. One day, however, I got a call from a woman who said she represented the Colombian government. They were going to hold an emerald symposium in Bogotá and would like me to come and give a presentation on our emeralds, along with a number of other experts in the field.

"Are you crazy?" I answered. "Do you know what I do, what I sell? I wouldn't last five minutes in Bogotá. No, I will not come," I said. I was unaware that they had any knowledge of lab-grown emeralds and was afraid for my safety. The reputation of lawlessness in Colombia was a well-known fact and being a direct competitor of natural emerald did not make me feel loved.

"Mr. Chatham, we know what you do, and we think it would be good for the local emerald dealers to also be aware of what you do," she said. "We will provide first class flights, a hotel and security. You will be joined by other speakers, such as Bill Boyajian of the GIA and Gary Roskin from *JCK* magazine, and countless other respected experts in their field. The Colombian government will provide everyone with security, and we will also visit the mining areas, like the Muzo mine by private helicopter. Please reconsider."

With the prospect of visiting the world famous

{119}

Tom Chatham, soaking wet, in the bowels of the earth kneeling on a board to stay out of the water.

Muzo mine, I reluctantly agreed.

I flew to Miami, then on Colombian Airways to Bogotá, all first class. As we taxied up to the gate the pilot came on the speaker: "All passengers are to remain seated until told otherwise. Would Mr. Tom Chatham please come to the front of the plane after we stop."

I got lots of stares and didn't like being singled out, not on this trip. When I got to the front of the plane and showed my passport, the door opens and I see two military guards, each armed with a machine gun.

"Follow us, please," they said.

So, with one in front of me and one in back, we walk right past customs. My luggage is already off the plane, and they put me in a Brinks-like armored truck. I think to myself, well, if they were going to kill me, they wouldn't need a Brinks truck, so I relaxed a little. We arrived at the hotel in Bogotá 30 minutes later and they whisked me directly to my room.

"Don't leave the room, a guard will be outside," they said. "We will call you when it is time to come downstairs."

About an hour later I get a call, "Mr. Chatham, you may come down to the main lobby and join your associates for a welcome reception. The guard will assist you down."

So off to party I went, with my guard. I knew about half the people there from all over the world: some from Brazil, Germany and Africa, as well as many from the United States. In addition, there were paramilitary types, all with an automatic pistol on their hip. These were the warlords and their bodyguards. Each warlord had a specific color guard in uniform, but the "boss" just wore a suit and tie. They were the important emerald miners and dealers in Colombia.

I spoke no Spanish and they spoke no English, but translators mingled throughout. When I was introduced to one of the warlords/dealers, they asked about my connection to the emerald business. I pulled out some crystals from my pocket and said, "I grow emeralds."

A whole lot of Spanish flew back and forth, and the interpreter asked for a clarification: "What exactly do you do?"

"I grow emerald crystals," I answered. "My name is Chatham. Have you never heard of Chatham Created Emerald?"

"No, we have never heard of this, is it new?" they responded.

"No," I said. "It is about 75 years old and my father invented the process."

And so it went all night.

One guy asked for a private meeting, with his guards, of course, and his interpreter. I displayed a number of uncut crystals, 5 to 50 carats each, on the table in front of me. In natural emerald this would have been worth millions.

"Mr. Chatham, you must keep this a secret, tell no one," the warlord boss said through the interpreter to me, with his paramilitary guards at his side.

I was finally able to get him to understand that our company had been growing emerald crystals in the United States for a long time, but that we did not sell into South America. This did not mean, however, that our emeralds had not been brought in by someone else. This was an interesting conversation, to say the least.

I was escorted back to my room after dinner and told not to leave until called. A guard was outside my door 24/7.

The next day was the conference, which was filmed and televised. I gave a talk on how to permanently enhance emerald using the hydrothermal method of growing emerald. I showed slides of colorless beryl crystals with cracks healed by green emerald. I don't think anyone understood what I was talking about. The Colombian audience knew nothing about crystal growth, so why should they understand my presentation? I also volunteered that they would have to figure out what this process would be called if they adopted it because they would be combining a laboratory-created growth into a natural stone.

This is done today with rubies in Bangkok. Heat-treating rubies at 1,000 Celsius allows some of the surrounding alumina to creep into a stone's cracks and actually "heals" the crack. They don't use the word "synthetic" to describe this process, just "heat-treated." The stone is covered in alumina powder to insulate it from overheating. Ironically, alumina is the main ingredient of corundum, or ruby. This was an accident with tremendous consequences; what should this new process be called? Naturally, the dealers just called it "heat treatment" with no mention of inducing crystal growth inside the stone. So much for equal treatment in gemology.

The same can be done with emeralds instead of just using oil to camouflage cracks, which is not permanent. To date, no one I know has used my presentation idea, but it is a good one because it is permanent.

Each presenter was given about 45 minutes and then time for Q&A. At the end of the third or fourth presentation there was an announcement for everyone to check their valuables. Someone was stealing the speakers' belongings or briefcases during their talks. The warlords were annoyed that someone would have the gall to come into this guarded arena filled with paramilitary types and commit petty theft. Guns were drawn.

Clearly, everybody was nervous, so one of the government organizers got on the microphone and told everyone to settle down, that everything would be taken care of. People lost their passports, plane tickets, money and personal items. The warlords were quite upset and put the word out on the street that everything must be returned…or else. I don't know how they did it, but the next day everything was returned to the hotel.

Now to the fun part, our reward, visiting the famous mines of Colombia, still with our military guards, of course.

Many stones were examined and traded in nearby villages, such as these two rough stones of exceptional color and quality. Cash was king and everyone was armed.

The Muzo mine was about 200 miles away from Bogotá, deep in the jungle with no real roads to get to it. An old Russian helicopter with bench seating on the walls got about 20 of us to the mine. It rattled and shook like crazy and since its maximum altitude was only about 5,000 feet, it couldn't go over the mountains but had to weave through them.

The mine usually has about 2,000 freelance miners working over the tailings, which were shoved over a cliff away from the main mine shaft. Everywhere you looked you could see small little emerald crystals, not big enough to cut and very included. It was entirely deserted the day we visited because the government got everyone out of the area for our protection. We later learned that every week four or five people get killed by other diggers over some disagreement. If you're out in this country, alone, you either have stones or cash in your pockets, so no one travels alone or unarmed.

I had never been in a real mine before, so I did not know what to expect. I soon realized it was a good thing I was not claustrophobic. We got into an open walled platform that held about 10 of us and slowly descended over a quarter of a mile down. Once we hit bottom, it was stifling hot and humid. Ventilation pumped in through large plastic tubes and there was some lighting, but it was pretty primitive. We were told not to wonder around or you could easily get lost. No one moved a muscle.

Once assembled, we were split up into groups to follow a guide for a tour of the mine. Shafts went up and down, some only 5 feet high, some split many times. The walls were wet and black and we stood in 2 inches of water. When we finally reached the end of one shaft, we saw a miner using a small jackhammer to dig into the shale walls, looking for a green hint of emerald.

We were each given a turn at the jackhammer, but when my turn came, I slipped a large Chatham emerald crystal in my hand. I pulled the jackhammer back and held the crystal up. I thought the head miner was going to have a heart attack, until someone quickly told him it was a joke, not a natural emerald I had dug out. That was the last time I pulled out any crystals.

Some people were absolutely paralyzed with fear at those depths, but for some reason it didn't affect me. Some rather important people refused to go down and they missed an experience of a lifetime. The Muzo mine is never open to tours or visitors.

Next, we went into the Coscuez mine, which did not go down in the mountain but rather went about a mile into the side of a mountain. Again, it was hot, and we were told not to wonder off or leave the group since one could get lost easily in these tunnels. Once out, we were all filthy and covered with black shale. The TV crews had arrived and were filming the mine owner and some of the dignitaries. About a year later, I heard this mine owner was murdered and his concession was taken over by someone else. Nobody was charged.

One of the U.S. magazine photographers took pictures of me picking up crystals in front of this mine owner and ran it in a magazine. "Chatham sure has balls stealing emeralds right in front of the owner," the caption read. Trouble was, I wasn't after emerald, but the quartz and iron pyrite crystals, which were very abundant in the Colombian mines. I was still just a rockhound at heart and couldn't ignore these beautiful crystals just laying around. The presence of quartz indicated silica, an important component of emerald. The iron pyrite was just a byproduct of the geological dyke that produced the emerald-bearing deposits. Iron also contributes to the color of emerald in a small way; chromium is chiefly responsible for the green color of emerald.

Then it was time to return to the hotel. We were given a choice, the old Russian helicopter or a small, two-engine prop plane. I took the plane, but unfortunately it also only had a ceiling height of about 10,000 feet and it was totally foggy down to 2,000 feet. We had a clear view through the open cockpit – and you could not see anything in front of the plane. The pilot was dodging this way, that way, up and down. I was extremely panicked and was sure we were going to run right into one of the mountains.

I am not afraid of rough flights, but this situation terrified me. This was a two-Valium, 20-minute trip, with Bill Boyajian next to me. At last, a runway appeared, and we set down, hard. I was never so relieved. These planes had no radar and flew by sight only, but since it was totally fogged in, I would guess that at times we were just 1,000 feet off the ground, trying to see under the fog while we zigged around mountains.

Back to the hotel we went, but all the guards were gone!

So I dressed down, no jewelry, Levis, T-shirt and tennis shoes, and several of us went out exploring the town. Bogotá was then a nice little town but is now much bigger. That night we had a big party at a local hangout, had too much to drink and had a good time. Everyone was glad they came – and glad it was over!

{123}

23
- - -
Finally, Chatham Created Gems, Inc. is Independent.

"Good job, Tom, we will celebrate when I get out of this hospital," [my father] told me. He never did, dying that night in his sleep. It was the Fourth of July, his favorite holiday because of the fireworks ... pure inorganic chemistry in action for Carroll Chatham.

Business was booming back in the states and so was my relationship with Jerry Hulse, who was still working for Heller Hope as the Chatham rep for the East Coast. George Heller, owner of Heller Hope in New York, wanted to retire. He sold his interests in the business to Golay Buchel of Lausanne, Switzerland, the cutters I fired for poor workmanship and doubling prices on me years ago. They were a big trading company and ran it that way.

I ended up in partnership with them because of the Heller sale and didn't like it. During one of the many trade shows with Jerry, (we were doing more than 20 a year) I discussed my displeasure with the new ownership and was thinking of opening a New York office if I could get out of my contract with Heller Hope. The contract had certain provisions for cancellations and one of those was a minimum dollar turnover per year of product they bought from us, with a 90-day warning of pending cancellation. Dave Heller, George's son, was now the manager. He had the option of "buying up" to the required level or quit.

I told Jerry they would not make the minimum this year and that I had plans to cancel the contract. I asked Jerry if he would join me if I did so but said there might be some legal problems if I tried to hire away their salesman.

"Tom, if you cancel that contract, they will fire me immediately. I don't fit in with them," he said.

So, I told him that if they fired him, he would be on my payroll immediately and that "we can begin the process of building Chatham New York." My father was extremely patient with me and told me that if I felt confident enough to open a New York office, he was all for it.

I sent a legal notice to Dave Heller of Heller Hope informing them of my exercising the cancellation clause due to their short falls in sales. I also didn't want to buy back their inventory, adding fuel to the fire. Dave Heller was very upset, and he told me so, but I explained that I did not want to be in bed with Golay Buchel, their new owner.

The first thing Dave did was to fire Jerry. Off the legal hook, I hired him and told him to find office space, start looking for people and enjoy himself for 90 days.

Soon after, we found a home for the company on 42nd St. and Madison Ave. in New York City. It was not in the center of the jewelry district, but close enough. Chatham Created Gems, Inc. was finally where it belonged, in New York, owned and run by Chatham.

Jerry Hulse knew everybody in the jewelry industry and soon we were back in business. Luckily, all the trade show contracts were in Chatham's name, so we didn't miss a beat because I originated all the trade show contracts, not Heller Hope.

We were getting bigger and bigger contracts from the "group of five" as I called them. This group represented the big manufacturing houses in New York and New Jersey that supplied 90 percent of the jewelry chain stores inventory.

All the big retail chains, such as Service Merchandise, Best Products, Zale's, etc., had buyers who supposedly made decisions on what they would sell in their stores. In reality, the big manufacturers led these buyers to believe what was best to sell – whatever the big manufacturer wanted to sell. If it was in a Best Products catalog, then by golly it had to be good, so we want the same thing but a little different, in our stores.

Our business grew from selling one stone here and one stone there, to orders for 5,000 pieces of 7x5 mm pear shape rubies or 10,000 pieces of 7x5 emerald cut emeralds to the big chain jewelry stores. This was completely impossible – then and now – in the natural stone world.

The New York office soon outgrew itself and we moved to spacious offices on 6th Ave. near 47th St. Deliveries were made on foot to the big manufacturers daily. We hired a 6' 4" and 220 lbs. "runner" who we thought would be safe from being held up.

Unfortunately, one day he was stopped at gunpoint on 46th Street, with a barrel in his back and one in the stomach. "Hand over

the bag" was the order, which he intelligently did. He came back to the office crying. He felt terrible and we told him he did the right thing, that it was not worth dying over and that his job was safe. These guys are going to be upset when they opened the package because of what they would find: not gold jewelry with diamonds they could fence, but Chatham Created Emeralds they could not sell, I told him.

We then began to ship everything by Federal Express, but it also had problems, losing entire trucks working within our zip code, 10036. FedEx switched to unmarked trucks and when that didn't work, it would not ship or pick up from that area code. FedEx created a new zip code for jewelry manufacturers and stone merchants at a new location just blocks away.

My father was slowly pulling away from the daily routine of the labs because he and my brother John did not see eye-to-eye on many issues, especially since John had begun to experiment with drugs and alcohol.

My brother became quite eccentric and wanted to only work at night, leaving daily instructions for his employees to follow the next day. He would go through periods of being completely clean of drugs to over-the-top hallucinations and mindless demands, leaving weird messages on our corporate answering machines. Our relationship also began to quickly deteriorate.

John was three years older than me, and I think he resented my leading the company. He was being very well compensated, but every year I would have to bail him out at tax time. He was also running the production side of the business into the ground and would have bankrupted us if not stopped soon. I was going to have some difficult decisions to make regarding his role and ownership in the company.

Carroll Chatham holding one of his emerald clusters we became so identified with over the years.

"Carroll Chatham was one of the truly remarkable men in the jewelry industry. He was a man of strong convictions, whose innovative intelligence and integrity contributed to the advancement of his field. He will be missed."

From an editorial by Richard T. Liddicoat on Carroll Chatham's passing, *Gems & Gemology,* Fall 1983.

John and I were in a partnership, but my father owned all the assets. He too was not happy with the way John was acting, but rather than firing him he asked us to buy him out. This was not my ideal solution to the problem, but I went ahead with it any way and arranged for a $1 million dollar pay out to my father. Thereafter, John and I were equal partners in the production labs as well as the marketing corporation.

Another concern was my father's health. He was deteriorating at only age 67. I privately asked a doctor what was wrong with him. "He is dying, but for no apparent reason," was the response.

My father passed away in 1983 at age 68. We never found out what he died of, but speculation from one doctor was that his exposure to all the chemicals he was around over his lifetime was just too much for his system. I saw him the night before he died, and he thanked me for all the work I was doing.

"Good job, Tom, we will celebrate when I get out of this hospital," he told me. He never did, dying that night in his sleep. It was the Fourth of July, his favorite holiday because of the fireworks. He always enjoyed explaining to me and John all the chemicals required to produce the colors of the exploding shells. This was pure inorganic chemistry in action for Carroll Chatham and we never missed the opportunity to go and watch the Fourth of July fireworks.

Our business grew from selling one stone here and there to orders for 5,000 pieces of 7x5 mm pear shaped rubies or 10,000 pieces of 7x5 emerald cut emeralds (such as this one) to the big chain jewelry stores. Chatham Lab-Grown Emerald. Credit: Orasa Weldon.

Padparadscha sapphire, another product from Chatham, in the 1980s. Once ruby (which is corundum, the same family as sapphire) was accomplished, we continued our research in all colors of sapphire. Credit: Orasa Weldon.

My father's passing did not solve the problems with my brother John, who became even more eccentric. I think he was always trying to prove his worthiness to my father and after his death could not do so. Finally, I became fed up with John's antics and asked for a "divorce." In this process, nobody gets paid, bank accounts are frozen, and a judge decides who gets what. Either John had to buy me out or I had to buy him out.

The court ordered an independent audit of all our operations and a dollar value was determined. I chose to buy John out. I would keep the corporation, the brand name, the sales offices and the right to produce any type of gemstone independently. John kept the labs, the processes, the Chatham brand name, and I would write a sizable check to him.

John could sell to anyone he wished, but always failed. I offered to buy crystals, but they were always too expensive, so I had to replicate what my father had built but in other countries. This was always a bone of contention because I had already begun to grow crystals in Greece, Japan and Russia, at far cheaper prices than our own production. I even offered John $500,000 per year to just quit growing anything, but he would have nothing to do with it.

John tried to sell to our customers but did not have the products they needed, cut stones. He fired most of his employees after about a year, including his son, Keith, and then closed down completely. All production was stopped; equipment was sold off or scraped. Fifty years of history went down the drain with not one memento to show for it.

He sold off all the property and platinum, moved to the Philippines with millions of dollars, and died shortly thereafter due to drugs, flat broke. He was 73 years old when he died in 2016. I flew to the Philippines to arrange his affairs and take care of his remains. I never found out where all the money went.

24
The Emerald Wars

"You have taken some million-dollar customers down to $500,000 without selling one more stone. These buyers are using you, playing one against the other. You will not get it back, you will not reach a point when Chatham is no longer there that you can raise prices, they won't let you."

– Tom Chatham

We were becoming overwhelmed with orders in 1993-95. It wasn't that we couldn't produce the crystals, but the finished stones were just not getting to the customers fast enough. When you're striving for quality in cutting and meeting stringent standards set by the buyers, it was easy for a few competitors to slip in.

Russia began to grow emeralds after the breakup of the Soviet Union and sold them cheap. The Inamori Created Emerald, by Kyocera, from Japan, began to explore the wholesale business from their Rodeo Drive retail store which, according to the people who ran the store, never made a profit. Biron Created Emerald from Perth, Australia came online with their Hydrothermal Emerald. Gilson, now owned by a Japanese company called Earth Chemical, was also entering the fray.

Each and every one of these companies, either directly or through distributors, began to knock on our customers' doors with a mantra that went like this: "Whatever Chatham is charging you, we will do it for 10 percent less."

And so it went, around and around, causing confusion in the marketplace. I called for a truce and invited all of the companies to a meeting in Bangkok during one of the trade shows we all attended. We all knew each other so I explained to them that they were killing themselves with this price war.

"You have taken some million-dollar customers down to $500,000 without selling one more stone," I said. "These buyers are using you, playing one against the other. You will not get it back, you will not reach a point when Chatham is no longer there that you can raise prices, they won't let you."

They didn't believe me – or didn't care. One rumor was that I was scared and couldn't stand the heat. Not true, I just couldn't stand to sell at less than cost.

It took me years to find out which company was doing the under cutting and in hindsight it made the most sense. Biron was represented by a big diamond dealer in New York and worked strictly on a percentage of the sale price. Since they were also selling the big companies their diamonds, they gave them the emeralds almost at cost. Of course, this didn't help Biron at all, and they eventually closed the production facilities in Perth, Australia, leaving unpaid debt to the originator, Artur Birkner.

We encouraged competition, such as this advertisement in the *The Morning Call* in September 1985 for Biron created emeralds, because it made more people aware of laboratory gemstones. Eventually there were about a dozen companies worldwide growing different gemstones all by varying methods. Today, only a few still exist.

Wholesale prices fell from $100 per carat to $20 per carat. Chatham did not and could not sell for this price, no matter what the volume. So, a finished piece of jewelry that was mounted with a Chatham stone went from $500 to the consumer to $139, maintaining the retailer's standard percentage markup. The trouble was that 30 percent of $500 is entirely different than 30 percent of $139.

It didn't take long for the big box stores to become disenchanted with the created stone business. They were also facing some legal problems because they were calling all these stones, Chatham Created. I had to threaten legal action against some companies because of flagrant trademark infringement. Finally, we stopped selling to these big box companies, way before prices reached rock bottom. That took about four years, from 1993 to 1997.

This price drop really exposed the cheapness of working in Siberia and the hydrothermal process for growing emeralds. In ruby, the manufacturers just switched to flame fusion ruby and sapphire with a cost

At one point we were very close to joining forces with the Biron company located in Perth, Australia, which I visited in 1996. The emerald price wars ended that venture, however and both companies continued separately. The Biron company eventually closed.

of 10 cents per carat. Since GIA and the FTC looked at all these products as just another "synthetic," they were on legal ground, but the product looked entirely different, and sales declined dramatically.

Some products, like Kyocera's Inamori, just ended and never sold created emeralds in the United States again. Some, like Biron, sold out, then closed. Some of the Russians got caught not paying taxes and were shut down, while one, Tairus, based in Bangkok and producing in Siberia, hung on, but to what? They were probably making $1 per carat on sales and only because cutting in Thailand and labor in Siberia was so cheap.

Chatham bowed out of all sales to big box stores like Costco, Walmart, Finley, Signet, Zales, etc. way before these price levels were hit, as did Gilson led by Earth Chemical and my good friend Kiyoshi Kobayashi from Kobe, Japan. Our volume was hit hard, maybe by 50 percent, but our bottom line went up. We went back to selling to the independent jewelry stores, our original core customer base created over the past 50 years. To this day, we will not consider selling to any big box stores under any conditions. Often, the buyers stop by our booth at trade shows, and we tell them to just keep walking.

I have come to realize that once a jewelry store chain reaches a certain size, new layers of management come in that have little knowledge of the jewelry industry and its traditions. I call them bean counters because all they cared about was how many dollars were grossed every year. The bigger they got the worse they got. I have never regretted walking away from that business.

25

Gem Trade Lawsuits

Lots of people sue big retailers for a variety of reasons, and they fight back hard, but when someone steals your name, you have no choice but to go after them. This puts the very credibility and reputation of our company on the line. I was not going to let all our talent and ingenuity fall to their shady maneuvers.

The United States is not as homogeneous as one might think. I soon found out that people in the state of Washington were not like people in Florida, and people in South Dakota were not like people in New York. In the mid 1990s, I was getting more and more calls from retailers who wanted to switch the offices they dealt with for our stones. Some just didn't want to deal with New York's way of doing business. New Yorkers always seemed in a hurry, or perhaps sounded a little condescending. Whatever the reason, it didn't sit well with many small retail stores in the middle of the U.S.

I drafted my brother-in-law Jim Starnes and his wife Cindy to open an office in Dallas, where they already lived. Jim was a retired Yellow Page supervisor and knew how to sell and manage the people he hired to answer phones. Cindy, my wife Dianna's younger sister, was well organized and kept all parts flowing well. The addition of the Dallas team meant that Chatham had three sets of inventories to oversee, a formidable task that was the responsibility of the San Francisco crew. Everything originated from San Francisco and was then shipped to Dallas and New York based on their needs. Another plus to this mid-America office was the ability to deal with the different time zones.

We were doing more trade shows and needed help. Jim had the dubious honor of coming on just when I sued the American Gem Trade Association (AGTA) for restraint of trade. They would not let us exhibit in their Tucson show, the largest colored stone trade show in the United States, because we were not selling a natural product. However, cultured pearl dealers were there, which are also considered a man-made gem. I asked again and again to be allowed to exhibit and was repeatedly turned down.

By this time, we had expanded our repertoire to about nine different stones. Emerald in two shades, ruby, blue sapphire, orange sapphire, white sapphire, yellow sapphire, alexandrite and opal, the latter coming from our new joint ventures with Biron and Gilson after the emerald wars ended.

I knew I had to do something about our New York office. Jerry Hulse was hitting 65 and was over his head dealing with these giants of the industry. Before they were his friends, but when the business became significant, as in millions of dollars, they demanded better prices, in-store displays and training for salespeople – areas beyond Jerry's expertise and abilities.

So, I decided to bring in a marketing manager to take the heat off Jerry, who knew nothing about marketing. I knew this new guy through industry contacts. He was young, aggressive and seemed to know how to handle these New York bullies. His main responsibility was to create ads and work with the agency we hired. Retailers began to call in more frequently and business was booming. Thousands of stones were being shipped monthly and I felt it was time for a change in leadership in the New York office.

The New York team was not happy with my choice of this new employee, who was the stereotypical brash in-your-face New York type and did not fit well with Jerry and the rest of the group. I had to make the hardest decision I ever made, putting the business and its continuation first. Jerry was the old-school version of the jewelry industry, while I felt this new person was the future.

Unfortunately, I had to let Jerry go over the phone. I had known him for 30-plus years, and he was a real friend, but I knew he was very unhappy. We didn't have 401K or retirement plans back then; when you left Chatham, you took your coat and left. In Jerry's case, however, I made an exception. "Jerry, this is what I am willing to do. I will pay you $2,000 a month as a consultant so your Social Security is not affected and will do so until the day you die." He asked me if I could put that in writing, but I couldn't (for legal reasons) and asked him to trust me. For the next 15 years or so he got a check for $2,000 every month until the day he died. We would visit during my trips to New York, and he seemed relieved to be away from the pressure and understood why I had to make the changes I did.

The new guy, no longer hindered by Jerry, was going full blast with all the major jewelry manufacturers. He was a collector of funds from all the big buyers who were selling our stones like crazy, but were slow to pay their bills. It always irked me that a company would order thousands of stones, mount and sell them all, then refuse to pay a debt owed, leaving it to us to chase them down.

For instance, we would fly to one corporate headquarters located in Dallas (with about 500 stores in the United States), with a six-inch stack of invoices going back six months, unpaid. We had to threaten legal action to get paid – they eventually all did, but it was like pulling teeth!

One time, the buyers offered us some bagels and announced, "Kimberly Emeralds sends these to us every morning, fresh!" Kimberly, a desk inside a large diamond supplier to this customer, represented Biron in Australia and they were killing us. I found later that they were the ones who kept lowering prices until we left the big chain retailers.

One of the first solo chores I had this marketing manager do was to call on a huge manufacturer a few blocks away in New York with a stack of invoices totaling more than $1 million that were a year old. The president agreed to see him expecting to meet and greet the new guy. The new guy dumped the invoices on his desk and demanded payment or a lawsuit was eminent. This guy went ballistic, "How dare you come in here demanding money, the gall of you!" Their attorney was also in the meeting and said, "Do we owe them or not?" That didn't make any difference that day and they threw this new guy out of the office. But we got paid the next week.

For a few years all went well. The new guy went overseas with me and saw all the operations that we had built up over the years. We had about 250 cutters in Shenzhen, China, as well as another factory for making jewelry. We also met with the Gilson and Inamori

{135}

Company heads we did business with before the emerald wars broke out. They were both in need of our created gemstones that they did not produce.

I shared everything with the new guy, a big mistake. Meanwhile, he was demanding more and more in salary and benefits. Unfortunately, his marketing ideas were not working well, and we were losing customers to cheap Russian goods coming in through Bangkok.

The industry thought there was no difference between a cheap flame fusion ruby at 10 cents a carat and a Chatham Created Ruby at $100 per carat. While gemologically similar, the two were visually miles apart. Both could be legally called "synthetic ruby," and many were switched. On the emerald side, it was Chatham at $100 per carat or Biron at $75 per carat; Inamori went to $50 per carat. Tairus, based in Bangkok, went down to $25 per carat.

And so it went, all the way down to $5 per carat for Russian goods cut in Bangkok. Of course, these stones were ugly and poorly cut so sales fell at retail. I asked one of the huge retail customers many years after, why a company making 30 percent on a $400 piece of jewelry, would switch to a cheaper stone with cheaper mounting, and sell it for $139 and still made 30 percent – but 30 percent of what? It didn't make sense. The answer I got from a person who ran a billion-dollar chain that went broke: "That's just the way the industry works, Tom, competition drives the price down until it's not worth selling."

The only formidable competition we had was Tairus, based in Bangkok and owned by Walter Barshai of Novosibirsk, Siberia, who knew he had his hands full. Profit was near zero, but he slowly did inch his way back up – although never to our former and present heights. That is hard to do from Bangkok, where all the manufacturers talk to each other and see prices of goods shipped in. So even though a manufacturer made jewelry for Zales, they also made jewelry for Costco and Walmart. There were no secrets in Bangkok.

One of the last big manufacturing deals I did in Bangkok was for an American retail chain of stores and involved the biggest stone supplier to Fabrikant, a huge manufacturing giant based in New York. They were also difficult to do business with, but we tried to, like everyone else.

When I met with them, Mr. Big Dealer says I can have all the created gemstone business if I meet a certain price. I said OK, I will meet that price under one condition: No one must know what that price is, and if word gets out, our deal is off. I simply cannot afford to sell everyone at that price because their volume is nowhere near yours.

Within one week I got a call from one of his competitors in Bangkok wanting the same price!

I called up Mr. Big Dealer and said, "I told you what would happen if word got out, and it already has. Our deal is off. I will not honor that price because you or your people can't keep a secret." Pissed is not even close to his reaction. "I will never do business with you again, you son of a bitch," blah, blah, blah and hung up. I haven't talked to him again to this day...and the other company did not get that price, either.

Chatham Created Gems stopped doing business with all the major manufacturers because of the insidious relationships they had with their suppliers. Some fired us for being too costly, some we just didn't like because of their attitude, suggesting; "Tom, you're too expensive, you should fire some of your people to lower overhead."

I could not and would not play this game and told all the majors to buy somewhere else.

The classic best and last example was with a chain store. An opening order with it can easily reach $1 million dollars, so we took the bait. As soon as our jewelry displays with advertising and in-store training was done, however, they copied all our designs. An identical, but cheaper-looking copy of the display was made, and all the jewelry was copied, but with lighter metals and an overall flimsy and unprofessional look, using cheap Russian created emeralds and flame fusion rubies. And, they put them both side by side, which made no sense!

The final straw was that they printed CHATHAM CREATED GEMS on the display. I told the buyer I was upset, and he told me to grow up. Well, I grew up. I gave them one quarter to get rid of all the Chatham displays and told them that no returns of product would be accepted. If this didn't happen, I would sue for trademark infringement. They complied and still ask to do business with us at trade shows. Sorry, I tell them, keep walking!

Copying jewelry designs is very common, unfortunately, but they crossed a line I couldn't tolerate. If something turns into a big seller, the chain stores will have it duplicated in Bangkok or China and tell you to just suck it up. Legal copyrights are rarely utilized because they are so hard to enforce.

Lots of people sue big retailers for a variety of reasons, and they fight back hard, but when someone steals your name, you have no choice but to go after them. This puts the very credibility and reputation of our company on the line. I was not going to let all our talent and ingenuity fall to their shady maneuvers. They paid up and dropped "the Chatham brand name" from their line.

Success breeds competition - and knockoffs. Some companies even use our trademark "Chatham" and to this day sell inferior gems and jewelry. Active vigilance is required on a constant basis. No other company in the world is allowed to use the term "Chatham" when describing any stone or finished jewelry that does not come from Chatham Inc.

We eventually fired all the chain stores who sold our brand name. When I got tired of being abused and mistreated, I switched to selling to the big suppliers that sold to the chain stores without the use of our brand name. This eventually failed as well.

We were going to concentrate on the independent jeweler, the mom-and-pop stores that were the foundation of the jewelry industry. The problem was my new marketing manager didn't know how to sell to them or what to sell them.

26
Re-evaluating Our Next Moves

Someone who helped me immensely during this time was Harry Stubbert. I am not quite sure when he came on full time, he sort of eased into the company… helping me in marketing, working from his home on a consulting basis… Before long, that turned into a fully salaried position. He was exactly who I needed to bounce ideas off in terms of marketing and human resource problems.

The demise of the new guy came about in 1996. He had a substantial contract, negotiated every year with our attorney, Bill Horwich, in San Francisco. Bill was no pushover and came with a history of litigation inside some of the most formidable legal firms in San Francisco. He once told me he had never met such a bulldog and their yearly squabbles over contract terms was annoying and tiring. This marketing guy never seemed satisfied.

When the emerald wars settled down and our business dropped in dollar volume, it was time to cut the cord and get rid of him. He was paid a set amount per year with bonus provisions, but without the sales to the big chain stores and their manufacturers, he was grossly overpaid and did not come up with any solutions to fix the problem.

He had negotiated a substantial severance package. By contract, I was required to give him a 90-day notice of termination, with pay. At the end of that period, I was required to pay a set severance amount based on time with the company, which I did. On my side of the contract, he was not allowed to conduct any competitive activities for a period of one year after our parting.

We all knew this was problematic in many states because everyone is entitled to earn a living, but there are specific exceptions that are not allowed and are legally binding. Among the areas that could not be violated were trade secrets learned at my expense, confidential business information, valuable personal information about the company, and significant relationships with current suppliers and customers. On top of this it was forbidden to take our list of customers. It can be very expensive to violate those statutes in any state. Employees have a right to work, but they can't do so at the detriment of their previous employer. This employee had confidential information that belonged to Chatham, and it was not his to use as he did.

David Heller, of Heller Hope, called me up after the termination. We hadn't talked for years following my cancellation of the contract between us. He was calling about this person and his background and I presumed he was asking for a job with Heller Hope, so I was not surprised by the call. I explained to David that I was paying him a salary of $250,000 a year by then, plus many perks, but he just was not earning his keep and was not solving the problems we were facing. I didn't care if he worked in the jewelry business, but he could not conspire to steal our customers and suppliers.

LAB-GROWN GEMS

CREATING NEW CHOICES

Specialized marketing is vital when it comes to moving lab-grown gems out of the showcase and into a customer's hands. Savvy retailers get this kind of on-target sales help from their suppliers of lab-grown colored stones, who show them how to guide customers in their colored stone choices.

By Keith Flamer

Many industry magazines around the world wrote articles and stories on how to sell a lab-grown gemstone. Our marketing position has always avoided any negativity towards natural gemstones and stressed the positive of existing alternatives … namely a Chatham laboratory-grown gemstone!

I never did find out if Dave Heller hired this person or helped him in any way, but it was suspicious. What I did know was that a Japanese firm we had been working with for some time was feeling the pain of our lack of purchases. Our slowdown in business with them was because we were in the middle of a huge price war and could not afford to continue buying from them. I don't know who called who, but Japan and the marketing guy got together in New York and conspired to undercut Chatham. What they did was unethical; what the marketing guy did was illegal and in violation of his contract.

I suddenly started to get calls from our biggest customers telling me they were being approached by a new company. The ex-employee contacted my overseas suppliers who

he met and did business with when he was working at Chatham, and suggested they fly to New York to discuss creating a direct competitor to Chatham. This was totally illegal on his part, and I could have put him in the poor house if I brought charges against him. As for the overseas suppliers and contacts, I just stopped doing business with them and told them why.

This ex-employee could have legally gone to work for a diamond dealer or jewelry manufacturer, even though it is in the same industry, but you can't take the customer list and all contacts and try to undermine your former employer's business. He violated every code of ethics in any job and especially one with a contract specifically forbidding it. In the end, he didn't pay the overseas suppliers and people lost their jobs because of it. The overseas firm eventually figured him out and begged me to come back to them and do business. I refused and the production division was closed and sold off – and I bought it for pennies on the dollar.

This guy's attempt to undermine our business failed and so did his company.

The New York office lease was running out. It had been moved to Long Island, because it was more convenient for the marketing guy and due to the type of customer we were selling to – independent retailers. Long Island, however, was not convenient to the jewelry business in Manhattan and I moved back into New York's central jewelry district at 47th St. in the old 580 Fifth Ave. jewelry building, right back where we started 40 years ago. I did not hire another manager but allowed the existing office personnel to run it on their own. "Make it work or I will close it down," I told them.

Meanwhile, Jim and Cindy Starnes in Dallas decided to retire and move to Florida, so I hired a manager to oversee the New York and Dallas offices. This worked out for a while, but I could sense we were losing ground. Working without direction and supervision may sound ideal, and can be, but it comes with the responsibility to be a self-starter and produce results.

The New York staff was goofing off, showing up late, covering for each other, while sales steadily fell. I am not a micro manager and allow people their space to do what must be done, but numbers speak for themselves. I soon became frustrated with performances in both Dallas and New York and made some major decisions. The manager in Dallas, partly responsible for New York's performance saw it coming. I called her up and said I was closing New York and she asked if

that included her too. "Yes," I answered. "It's just not working. Then I flew to New York unannounced and fired everyone on the spot. Nobody looked surprised.

I had just signed a five-year lease at 580 Fifth Ave. and still had four years to go. I told the building manager I was going to leave but would sublet the office to honor my lease. Landlords hate subleases, but they also hate chasing out of state tenants for rent payments. A sublease must be honored, or else they could just cancel the lease; it's their option.

The office was nice, newly painted and furnished with at least $40,000 in improvements. We had the best safe, good security features and even a man trap front door. I soon got a potential renter, and I asked him to just take over the lease and pay me $500 for everything in it, a super deal. The landlord jumped right in and in typical landlord fashion, took over the lease, which released me.

Dallas continued to run with a crew of two or three, but their numbers were becoming dismal as well. One guy was sleeping on the job at his desk if he was there, and some were sending out hundreds of stones on memo and not following up. Sales plummeted.

Someone who helped me immensely during this time was Harry Stubbert. I am not quite sure when he came on full time, he sort of eased into the company, contributing more and more from his home near Carlsbad, California. He had risen to a position of CEO of the instrument division of GIA but left to run a startup business installing jewelry kiosks in jewelry stores that let customers virtually try on jewelry. It was an idea way before its time and within a few years ran out of financing and closed, leaving Harry without a job.

Chatham's finished jewelry pieces showcase our lab-grown colored stones and diamonds and are very popular. We focus on desirable and affordable jewelry that is practical to produce. We never copy what sells in natural stone jewelry, instead striving to create styles not often found in natural stones, such as these onion-cut Alexandrite drop earrings. Innovation in cutting distinguishes our line from the multitude of other jewelry manufacturers that feature either natural or lab-grown gemstones.

Harry began helping me in marketing, working from his home on a consulting basis, while he contemplated his future. Before long, that turned into a fully salaried position. He was exactly who I needed to bounce ideas off in terms of marketing and human resource problems.

His first chore: the Dallas office. He and I flew to Dallas to assess the situation and we both agreed it was unsalvageable. It was near the end of its lease, which we did not renew, and the remaining employees were told of our plans to close the office.

Harry next helped me realize that it was time to think of new ways to sell Chatham Created gemstones. I knew the best way to control our destiny was to control our product. Since our end buyer was the retail store, perhaps we should sell directly to them – not with loose gemstones, but with our own finished jewelry line using our colored stones and later our diamonds. This would keep our brand name clean and get us as close to our customers as possible.

I found excellent manufacturers in Hong Kong and China, but I found out that you must tell them what to build. Every country has its own tastes. What sells in Great Britain won't sell in the United States. What sells in the U.S. won't sell in China. It is very difficult to identify each country's tastes. Our manufacturers produced great, well-made jewelry, but it did not have a "look" that captured the interests of the U.S. consumer.

Harry saw this immediately and we went to China to work with our manufacturers. They immediately sketched out the same looks as before: big, bold European styles that did well at Bulgari stores, but not with Chatham stones and U.S. buyers. We needed to keep the price points within reason and still make a desirable piece of jewelry.

This is probably the most difficult aspect of creating jewelry. Desirable, affordable jewelry that is practical to produce. Big box jewelry retailers and the people who design for them are geniuses at doing this.

So, with little help from me, Harry and the designers developed lines of jewelry using our unusual cuts, plus he came up with the marketing to sell it. I scoffed at the idea of minimum purchases for retailers, but Harry was right, and continues to be so.

Having a selection of one piece of jewelry or loose stone in a showcase has never worked and that's where we were: one stone here, one there, a piece of jewelry there, one here. It just doesn't add up to ongoing sales. Getting space in retail stores is tough and we all fight for every square foot of it. Harry suggested proper displays, generous return policies and exchange privileges. He took Chatham jewelry from a few inches of space to full counters of jewelry running six feet long – and selling well!

I would be the first to admit, I can do a lot of things well in growing crystals and cutting stones, but designing jewelry? I flunked! I was a disaster in that department.

27

Overcoming Disasters

Everyone was fine, but we soon realized we had problems in our labs because all the power and gas were shut down … We could stand about a four-hour outage in our crystal-growing furnaces – it would show in the finished crystals, but we would still get production … More than four hours and you will get all sorts of growth problems, like a cancer, with little crystals forming on the crystal faces that will not heal, no matter what you do. This totally ruins the run.

Speaking of disasters, we've had our share, which basically fall into two categories: natural disasters and incompetence from third-party suppliers.

The 1989 Loma Prieta Earthquake hit San Francisco at about 5 p.m., while the San Francisco Giants were playing the Oakland A's in the World Series. I was standing in my office doorway when it hit. Since I was born and raised in San Francisco, I was used to earthquakes and normally marveled at the power displayed as the roads waved and bent up and down and streetcar tracks wiggled around like snakes. It was cool to watch and see – if you were out of harm's way. None had scared me before or had done any substantial damage in my lifetime.

On this afternoon in 1989, however, I immediately knew what was happening and grabbed a door frame, which was just about useless in our commercial office building with glass walls. This was a powerful earthquake, and I was not amused or full of scientific awe. It was scary.

Seconds later, we heard a tremendous crash as about 25 windows – about 10-foot square – popped out of their frames in the I. Magnin building at Geary Street and Stockton Street, which was 100 yards from our building. It was a miracle no one was hit or injured, but others were not so lucky as one freeway pancaked down onto another, killing many. Most of San Francisco was saved from destruction because of the retro earthquake upgrades the state required.

Our office building was completely up to code with huge cross bars showing in every window. Everyone was fine, but we soon realized we had problems in our labs because all the power and gas were shut down, a protocol that was implemented after the 1906 "Big One" when fire, not earthquake damage, ruined much of the city. We could stand about a four-hour outage in our crystal-growing furnaces – it would show in the finished crystals, but we would still get production. When the power is cut off and temperatures drop, a fine line will be seen at the end of the run. More than four hours and you will get all sorts of growth problems, like a cancer, with little crystals forming on the crystal faces that will not heal, no matter what you do. This totally ruins the run.

I immediately called a generator company to rent equipment – we needed a generator that was big enough to require its own 18-wheel trailer. They were sold out and any that might have been available went to hospitals and urgent care places first.

The Loma Prieta earthquake hit San Francisco on Oct. 17, 1989. Although many learned the lessons taught by the 1906 earthquake, some buildings were not up to code and collapsed.

None of our employees could get home because they all lived in the East Bay, across the Bay Bridge, portions of which had collapsed, so we all jumped in our cars and headed out to the labs where I found John, just dumbstruck. There was nothing we could do for most of the furnaces, but we immediately began emptying those close to being finished so the crystals would not be damaged by the cooling flux. We saved maybe 12 out of 75 furnaces in production.

The rest got hard as a rock. When molten rocks cool, there is shrinkage, which is fine, but when you try to reheat the crucible, the material in it unevenly expands to its previous volume, which put the crucible at risk of bursting. Many did when we restarted them weeks later when the power came back on. We ended up with about a $2 million dollar loss, at minimum.

We had always known we were vulnerable to earthquakes, so our furnaces were bolted down, but we had done nothing to protect against a long-term power failure. I had tried to find insurance and equipment solutions, but nothing really worked for this situation. Earthquakes falls under the "act of God" clause, so insurance was worthless. I even explored business interruption insurance but could see no way that would apply either.

I considered buying a standby generator, but there was no place to store it. The roof was a possibility, but I knew the neighbors would have a fit because generators are only good if run at least once a month under full load. This might be tolerable in a disaster, especially if we shared the power with others around us, but this was not realistic to do on a monthly basis because of the noise it creates. Then there is a fuel storage problem: diesel needs to be used constantly and be safely stored, or it rots.

We never resolved this dilemma. We added up all the years of having nothing to protect us against the likelihood of it happening again and continued my father's choice of being self-insured, crossing our fingers and hoping for the best.

Another one of our disasters was caused by incompetence in the early 1990s.

This disaster started in our ruby production lab when we noticed an electrical warning of a leak in one of our 50 furnaces. The warning meant the crucible was leaking and the molten flux was getting down to the element, which would burn out in a matter of hours, as the flux ate up the metal element. This was not a common problem, but then other units set up at the same time also failed, then another, and soon they all started failing.

We could see this by watching the level of electrical conductivity between the crucibles and the elements. The conductivity should be zero because of all the insulation around the crucible, which were sitting in clay muffles with a clay plate at the bottom. This was a very simple but effective alarm system. It wouldn't save the run, but it would maybe give you enough time to remove the growing crystals before they became etched by the cooling flux. We also can save expensive flux this way too. Each unit has a gauge above it showing the degree of conductivity, between the element and the crucible, which is usually about zero.

By this time every crucible was failing. We immediately emptied each one and called our supplier, Engelhard Industries in New Jersey, where we purchased the pure platinum sheets our crucibles were made from, to report the problem. We sent samples of our crucibles to them for testing and noted where we saw the leak areas. Engelhard suggested that we had changed our flux material and were at fault, but we had changed nothing. Engelhard could find nothing wrong with their metal, but because we were a good customer of 50-plus years standing, they would credit our refining charges. But then what? We would get maybe $50,000 in a refining refund, but we lost more than $2 million in ruby production on top of that. We said no thanks on their offer and moved our account to Johnson Matthey, another big refiner.

We began to process all our metal through Johnson Matthey and the results were perfect, no leaks. They were very innovative and made many special parts for NASA as well as the medical industry, such as platinum stents for heart issues. We were back in full-time production in the ruby division within four months. It was odd that this platinum metal issue never effected our emerald production since all the metal we used came from the same place. It turned out that we were just lucky.

A year after we had left Engelhard, we got a call from the head of Engelhard who requested a meeting. We agreed, but I knew we were not moving back to them. One of the top executives came in from New Jersey, hat in hand, and said they found the problem.

We used pure platinum, 99.999% pure. That means there is .0001% of some foreign element remaining and at high temperatures, our fluxes are extremely reactive with anything it

The I. Magnin building, located just down the street from our corporate offices, popped out more than half of its 10-foot windows, which shattered and rained on busy Geary Street below. Miraculously, no one was hit or injured, but many lost their lives on the freeways that pancaked on top of one another.

encounters. Pure platinum is inert and non-reactive with most other elements but not many other elements share this ability. So, what was the .0001% impurity made up of? This is something Engelhard should have thought of immediately and investigated when we had our issue, but it is no small feat to determine such a small amount. This is why I call it incompetence; Engelhard should have found the problem immediately.

When he came to see us, the Engelhard executive told us that they were able to zero in on one of the leak holes using an electron microscope and found a tiny crystal of Zirconia growing in it. This .0001% impurity of Zirconia loved our flux, began to grow some weird combination of elements and pushed a hole open as it grew to a few millimicrons in size. Our fluxes are extremely viscous at high temperature, so even though the hole was tiny, the flux went through it like a freight train in every ruby furnace!

{147}

The executive had a picture of the crystal in the hole that he gave to my father. He apologized on behalf of the company and asked for forgiveness and said the employee who had blown us off so quickly was no longer with Engelhard. He didn't ask for the account back because he knew it would cost at least a $2 million in repayment, but he did stress that their metal was now free of all contaminants and an ongoing testing procedure was in place and used on all metals being refined, especially for crystal growth. And crystal growth was becoming a huge business in Silicon Valley, so this occurrence had major implications.

We really didn't feel vindicated because we knew from the start the leaking issue was not our fault. A few weeks later I got a sales call from Engelhard and this person told me they were put in charge of getting back the Chatham account, come hell or high water. I told them Johnson Matthey was doing a great job, so if a $2 million dollar compensation was not in the equation, we were staying with them. They never called back.

Next, we faced another disaster due to problems caused by our electric supply company.

We expanded the square footage of the San Francisco lab in the early 1990s to double emerald production and upgraded our electrical capacity by 100 percent. We were anticipating a draw of 400 amps, so a massive electrical panel was installed to handle it. Everything from your panel to the service provider is their territory and only they can work on it. One of the rules we always had with them was that all work had to be done "hot" with no power shut off, so we could keep our furnaces running. You pay extra for this because it is highly dangerous and requires a bigger crew.

The electric service crew could see we were not drawing even close to the power we designed for, so they changed the incoming wire mate-rial to aluminum, which they thought would work. This immediately melted, however, and we insisted on copper wire of adequate size because we were ramping up in the near future – the whole point of increasing our electrical capacity. The wire size they installed to our building was not even half of what was required, 8-gauge instead of 4-gauge (smaller numbers mean bigger wires), but the job was inspected by the city and signed off on. Almost immediately the wires melted, the company came out and just reconnected the same size wire. Meanwhile, we are adding furnaces and electrical load.

During the night of Nov. 1, 1992, the neutral wire of our incoming power lines melted and caused a huge electrical surge of 440 volts of power. I arrived at the lab around 8 a.m. after our alarm company called and said the power was off and found a dozen guys working on rewiring our electrical panels with the correct 4-gauge wire. The street was covered with wire, and I was told not to touch anything; "This is our property," so I took pictures. They then took all the undersized wire back to their yard and destroyed it.

The correct size service wires may have been replaced, but everything in the lab was ruined, burned out by electrical overload. Brother John was responsible for figuring out what our losses were and what was usable to some degree from the furnaces. In addition to all the emerald lost, we lost 150 small computer temperature controllers at $5,000 each and every light bulb, computer and furnace – anything hooked up to electricity (turned on or not) – was fried. We sued them for damages, about $2 million dollars.

There is a tiny clause on many bills you receive that states that an arbitrator must be used for all disputes. So, we hired an attorney who specialized in arbitration and went to work digging up evidence. In the discovery stage of the lawsuit, they had to share their "trouble reports" with us and it was clear that we had had power problems reaching back 10 years with their service.

We had multiple failures, but not consequential enough to sue because only one portion of the service was melting and was easily repaired. Unfortunately, crystal growth was affected every time this happened. We had probably five melt downs in seven years, and they just kept putting the same gauge wire back up and not telling us that. We had zero control over what they did.

At the time, we were slowly adding four new furnaces a month, aiming for a total of about 100 units, and the furnaces are not "on" full bore every minute. The computer temperature controllers on a furnace senses a fall at a set temperature and turns on power, while the furnace next to it may be on the downside. These "coasting" spells could last minutes, depending on the efficiency of the insulation. The whole lab was like a gigantic slot machine with one or zero on each temperature controller. The odds of the controllers being all off or all on across the board at the same time were millions to one. We hit the electric jackpot on that November night when all our furnaces drew the full 400 amps, all at once.

We tracked down the original employee who put in the service, now retired and explained what happened and asked for his help with our lawsuit, for a fee, of course. His response: "Sure I remember that job. I would be happy to help!"

Litigation took about a year in front of the arbitrator, who found in our favor and ordered them to pay all costs, plus legal fees, for a total of about $1,834,000. We thought we had won, but they did not accept the decision of the arbitrator and we had to sue them in court.

Their attorneys and our attorney went back and forth for months. We scheduled a conference call to come up with a number we could all live with. Unfortunately, this was right in the beginning of my diamond research in Siberia. I don't know why we agreed to this time frame, but it seemed like it was this or nothing, so I said I would call from Siberia.

I found myself in the one hotel of Novosibirsk, Siberia. It was May, so it was not freezing, but the hotel was really a dump, with toilet paper shoved into the window cracks to keep out the cold and wind in the winter, a rust-stained toilet and moldy shower, and thread-bare rugs over cracked linoleum. The phone in the room was monitored by management and I was warned not to discuss anything in the room with a guest because it was bugged.

The time for the phone call came – my morning in Siberia and their night in San Francisco. The front desk put the call through to my room and the discussion started. Present in the conference room in San Francisco were the attorneys representing the service company, our attorneys and Dianna Chatham. Not five minutes passed and there is pounding at my door. I put the phone down and open the door and this Russian lady is yelling at me "rubles, rubles for phone call." I don't speak Russian and her English was almost zero, so I pull out my bill fold and show her many U.S. dollars. "OK, OK you pay when done," she tells me.

We made our deal with the service company (more of a take-it-or-leave-it offer) for about $900,000. We were spending about $30,000 in legal fees per month and if we went to trial it would balloon. I didn't like it but took the $900,000.

Then bang, bang, bang on the door again and I had to pay about $300 to the front desk lady for the privilege of the incoming phone call!

28

Diamond: The Holy Grail of Crystal Growth

"Because we know how to grow large crystals better than any other company in the world. I have been in the created gem business for more than 60 years and we have long-standing relationships in the jewelry industry. People will want this new product. Growing gem diamond is the holy grail of gemstones."

– *Tom Chatham*

My father had been interested in growing diamond since he was a young boy (remember his Henri Moissan experiment?) and I was also fascinated by the concept, so I began to explore who, if anyone, was making progress in this created gem.

I read an article in *Gems & Gemology*, GIA's scientific journal, about a created diamond made in Japan that was being used as a heat sink. Diamond has a unique ability to conduct heat, very quickly, and then dissipate it in the air. This quality is very important around computers, which, at the time, had to reside in air-conditioned rooms to survive. The computing methods used then created a lot of heat.

The company was making little radiators out of diamonds they made. I reached one of their United States outlets and bought a few of the diamond radiators or heat conductors, which were about 1/4-inch square and about 3/16-inch thick. The prices were very reasonable, but I wanted to cut them into gemstones, which amounted to a significant loss of yield but showed me it could be done. The stones were very clean and a bright canary yellow.

I called up the company headquarters to introduce myself and explain what Chatham Created Gems, Inc. did.

"I bought some of your diamonds and cut them into gemstones," I said. "I would like to talk to your company about a joint venture growing gem diamond rough suitable for jewelry use."

I thought they would be happy to find a new use for their product, but quite the opposite happened.

"We do not want our material in the gemstone business, and you are further barred from buying from us ever again," was the response I got.

I was shocked and bewildered. "Why don't you want to expand the use for your diamond process?" I asked. All he would tell me was the bosses in Japan forbid it, end of story.

"OK," I said, "but do me this one favor. Pass on my name and background information to whoever calls the shots in Japan and ask if I can meet with them, I go to Japan frequently."

He agreed, but added "If you hear nothing back, don't call again."

Many Chatham lab-grown diamonds were requested for further study by GIA, such as this 0.63 carat yellow crystal. Dr. James Shigley, distinguished research fellow, wrote a lengthy report on it in the Summer 2004 issue of *Gems & Gemology*, the prestigious scientific journal produced by GIA. Credit: James E. Shigley/GIA.

I received a telegram from Japan 30 days later: Mr. Kuzomoto (not his real name) will see you at 1:30 p.m. on March 31. We are in Tokyo; tell us which hotel you are staying at, and we will send a car. Please confirm.

Wow. I was shocked and delighted!

So off to Japan I go with all sorts of brochures and samples of our work and anything I could think of to convince Mr. Kuzomoto we were on the up and up.

A big black limo picked me up at the hotel at the prearranged time. Typical Japanese style, the driver wore white gloves and a black suit and tie. He did not speak English. We pulled into a guarded complex 20 minutes later where I had to show my passport. The place was a huge, sprawling compound with a tall office building in the center.

I was shown in and taken to a board room containing a 40-foot-long table. Mr. Kuzomoto sat at one end with two scribes on either side of him. He spoke perfect English and welcomed me in with a bow and pointed to a chair.

"Mr. Chatham, please make your presentation," he said.

So, for the next 40 minutes I gave him the history of Chatham while the two scribes furiously wrote everything down. I explained our position in the industry, our integrity, my father's desire to make diamonds, and how this new application for their diamond process could reap huge profits for this company. It would be an amazing accomplishment to brag about.

"How much do you think you could sell, Mr. Chatham?" I was totally unprepared for this question and very intimidated but had to give some answer.

"Well, if we could produce the right sizes and color, I would guess anywhere from $30 to $50 million per year as a start. The natural di-

Chatham lab-grown diamonds include colorless, pink, blue and yellow - all grown by the HPHT process, except the colorless emerald cut that was developed by the CVD process. Sizes range from 0.33 carats to 1.54 carats. This image is featured on the cover of "Laboratory-Grown Diamonds," by Dusan Simic and Branko Deljanin. Credit: Sharrie Woodring.

amond business is about $20 billion per year, and I see no reason why we could not capture 5-8 percent of that." It was pure speculation on my part but not far off, I would later come to learn.

"Mr. Chatham, the company has thousands of products we sell very profitably all over the world. Why do you think you can do better than we can?" he asked me.

"Because we know how to grow large crystals better than any other company in the world," I told him. "I have been in the created gem business for more than 60 years and we have long-standing relationships in the jewelry industry. People will want this new product. Growing gem diamond is the holy grail of gemstones."

"We sell millions in emerald and ruby; I am confident diamond would be 100 times more successful. The majority of jewelry sales by the numbers involve a diamond. We could capture 15-20 percent of that. Your company, Mr. Kuzomoto, has no experience in the jewelry industry, which can be very difficult to break into. The biggest hurdle is one of resistance; people will not welcome you with open arms.

We, at Chatham Inc., have overcome that attitude to a large degree."

"Mr. Chatham, we have an unwritten agreement with major diamond producers. We will stay out of the gem business, and they will allow us to co-exist with them in the electronic business," he said. "If we break this promise, there could be huge financial repercussions for us. For this reason, we must turn down your offer of assistance. Thank you for coming, we will think this over."

In Japan this means no, don't call us, we will call you if there is any change.

This meeting really blew me over. I was granted an audience with a board member of one of Japan's biggest companies. I was surprised they would fear anyone due to their immense size and power, but scared they were. Later, they would feel that wrath of economic power, as many have experienced. Companies like De Beers have a 100-year reputation of ruthless competition and reprisals.

So, I went home to the United States and an ill-fated Russian-Ukraine endeavor.

29
Seeking Business Partners in Russia

A hydrothermal emerald of dubious quality [from Russia] … was cheap, about half our price … When everyone was comfortable, I asked to be introduced to the people in Russia, where this new emerald was grown, because I wanted to establish some sort of working relationship with them to address the confusion and price wars in the marketplace – the emerald wars.

Chatham Created Gems entered the created diamonds business by way of emeralds, which was fitting since this was how our business began. By the early 1990s we had three competitors in emerald: Gilson in Japan; Regency, a hydrothermal emerald from Union Carbide; and Biron in Perth, Australia. Then a newcomer from Russia came on the market that changed everything.

It was a hydrothermal emerald of dubious quality, but it was cheap, about half our price. I did my usual searches and found a United States distributor, Dan Kim, for the material and reached out to do some business with him. When everyone was comfortable, I asked to be introduced to the people in Russia, where this new emerald was grown, because I wanted to establish some sort of working relationship with them to address the confusion and price wars in the marketplace – the emerald wars. This was agreeable to Dan Kim, with a pre-agreed fee going back to him if anything was accomplished.

In early spring 1992-93, Dan Kim, his son, who spoke some Russian, and another interpreter flew with me to Moscow, a huge, cold, uninviting government city in a country that was breaking up and out of control. I stayed at a decent hotel for $400 per night, not the opulent Metropole Hotel at $600 per night. Why Russia was so expensive was beyond me, but it was obvious that foreign visitors were being overcharged.

The sun stayed at half-light until midnight, and it was freezing cold, but we ventured outside. Thousands of people milled around Red Square and the streets were lined with people trying to sell something, anything. The lines extended block after block: pickles, nuts, flowers … one lady was trying to sell a handful of parsley. One stand sold burned out light bulbs, which I later found out were used to replace good light bulbs that were stolen.

Someone overheard me speaking English and approached me to say, "This is all your fault, you have caused us much harm." We slid away because these people were serious, but I didn't know why. I later learned that these pensioners who lived in government houses and received about $15 a month, were starving and assumed the United States government's involvement in the breakup of their country was the cause of their struggles.

I took a picture of a guy selling a kitchen sink to prove the adage "everything, including the kitchen sink was being offered." I was about 200 feet away and used a long lens on my camera so the people I was

Citizens of Moscow lined the streets selling anything they could get their hands on (even the kitchen sink!) after the breakup of the Soviet Union in the early 1990s. They were not pleased with me photographing them because it was embarrassing, and I was punished with a policeman's club for taking this picture. I later found out that they thought I worked for a publication like *Time* magazine and that the photo would be on the cover.

This was how "retail" worked in Russia at the time. A truck carrying goods from the surrounding countryside, such as this load of potatoes, would pull off the street and open the back and people would run up and buy what they could carry.

photographing would not feel violated, but suddenly a cop hit my camera with his club and started yelling at me in Russian. I didn't speak a word of Russian, so I slowly unlatched the long lens and put it and my camera in my bag. I didn't know what else to do.

I said "American," and his voice got louder. I had little choice but to walk away and hope he wouldn't crush my skull with his nightstick. I slowly backed up, turned around and walked away. Nothing happened, so I walked faster and kept my head down. No more pictures.

Later that night I told my interpreter about the exchange and asked what he thought had happened.

"The people are embarrassed to be out begging for food and money and they have no idea who you are," he said. "Maybe they think you work for *Time* magazine and that shot will be on the cover with some humiliating title."

He was right, I was seeing the Russia few people from the West had ever seen at that time, and it was ugly. Russia may have been a superpower militarily, third behind the U.S. and China, but economically it was impoverished and starving. Communists don't believe in capitalism and think if everyone works hard and shares equally, they will all benefit. The trouble with that concept is it takes away incentive. Why work harder than your neighbor if there is no extra benefit? Why work overtime, with no reward? It reduces people to a common denominator and in Russia's case it was near the bottom.

Years later, on a trip to visit another diamond producing company in Saint Petersburg, I found that Russia had come full circle and appeared to be very robust with a thriving economy. Everyone loved what Putin had achieved ... back then.

One day on that trip we traveled four hours by car to a small research laboratory in Minsk that we heard was growing emerald crystals of many types. I was put up in a small hotel

{157}

where they took my passport, a common practice. I think the room cost $10 a night and felt like it. The bed was a cot, and the heat was minimal. I skipped the shower, it looked so dirty. The next morning when we tried to check out, a heated discussion was going on at the front desk. I asked my interpreter what the commotion was about, and he told me they wanted an extra $25 because I was an American, so I must be rich. They refused to return my passport.

"Let's go, we can deal with this later," my guide said.

"No, my passport, I need my passport," I said, but we left anyway. You are screwed if you don't have a passport in Russia, so I worried about it for hours until we arrived at the research center.

Back then, anything built in Russia looked run down even if it was brand new. Tiles were missing from the steps, the halls were dark (no light bulbs), the floors were bare concrete and there was no heat. We went to a conference room to meet with the director. Unbeknownst to me, the first discussion was over my passport. The director picked up his old Bakelite telephone, barked a few sentences and hung up.

We talked for a few hours concerning our needs and had lunch. The level of research I needed was not sufficient for diamond or emerald growth yet, but everything will "soon be better" I was often told. Then, out of the blue a man walked in and handed me my passport. The director apologized and said the hotel would be penalized and paid nothing for their rude actions. Cool. I met my first Russian muscle.

So, finding nothing of interest in Minsk or Belarus, we flew to Novosibirsk in Siberia (where I made that $300 phone call with the electric company's attorneys to settle our lawsuit). Russia is huge, mostly unpopulated and has 11 different time zones, so it was a 10-hour trip on Aeroflot, the national airline of Russia. The main airport in Moscow was a complete dump with garbage all over the place and very few people working. We stood in line while the attendant just kept talking to her buddy next door. Finally, my interpreter spoke up and asked for service. She reluctantly looked at our tickets and passports and dismissed us.

"What about our luggage?" I asked. "We take it on ourselves," the interpreter said. OK, that's a new one on me, I thought.

We went down to the tarmac to a 747 aircraft with its bottom belly door opened. We lined up to get on board, but when we reached the bottom of the plane, we were directed to a short staircase into the belly of the plane with our bags. Everyone just threw their luggage into a pile and proceeded up two flights of stairs inside the 747 to the first deck to pick out our seats — all economy class.

As we were walking around, the plane suddenly started to move — no announcements, no safety demonstrations — the pilot went down the runway then pulled straight back on the yoke. I opened a bottle of wine; this guy was flying a 747 like it was a MiG fighter plane. I was scared to death the whole way and the landing was just as severe. I never flew Aeroflot again.

The customs area in Novosibirsk, which was spun off from Russia like many other satellite countries, was just inside an outer wall of glass so everyone on the outside of the airport could see the arriving foreign passengers. We could see a bunch of guys, mostly taxi drivers, standing around a fire in a 50-gallon drum outside.

I had about $25,000 in cash on me so declared it. The agent asked to see it, so I showed her the four-inch stack of hundreds. The agent waved her hand and shouted something in Russian and I looked through the glass and saw that every eye around the fire was on my stack of cash. That was lot of money to anyone in Russia, the equivalent of $500,000 U.S. dollars. What a good way to get rolled, I thought. Another agent walked over, looked at the cash, wrote something down in a book and handed it back to me. I told my guys to go get the car and pick me up at the front door. I ran to the waiting car.

Huge research facilities with thousands of workers — many were used for military research — were abandoned when the Soviet Union broke up. They had the facilities, but no direction or idea how to make a living, including smart Ph.D.'s who could make products people wanted. Whole factories were given away to those who could pay the employees and create an outlet for their production. It was a free-for-all from 1990 to about 1995, and muscle played a big part of where you ended up. Many would make a fortune. That is where the oligarchs were born - and flourished.

These areas were previously off limits to most citizens, much less an American. Not being Russian, or able to speak Russian, or having any Russian connections pretty much shut us out of these opportunities. It was the Silicon Valley of Russia. Once spun off from Mother Russia, they were left to their own survival.

That's where making emeralds came in, however, and I was getting much closer to the source of my competition. There were huge opportunities here, if only I could pull them together. The Russian growers had such an advantage over me: no rent, no taxes, no one to answer to, no cost of goods. It was a license to steal, and some did. While I was there about

The Russian Academy of Science in Novosibirsk, Siberia, was the source of many tons of cheap hydrothermal emeralds, causing much disruption in the American market. They did not use any crucibles in the autoclave furnace, just a steel pipe that eroded in the super-hot solvents. The eroded metal became a part of the growing emerald crystals and created a unique "oil in water" inclusion. This method was extremely dangerous because the pipes would explode when heated to 500 Celsius and spray deadly beryllium throughout the labs.

600 ounces of platinum went "missing" and no one was caught.

Word of my arrival traveled fast in the scientific circles, and many knew of my father's successes. I was warmly greeted by many scientists growing emerald crystals who insisted on calling me Dr. Chatham — until I asked them to stop because I didn't have a Ph.D. Of course, everybody there was a Ph.D. because education was free in Russia until you hit the limit of your intellect. Once you failed to move forward, you were assigned some menial task or labor job, but many of the researchers I met in this facility had multiple degrees.

One of the first labs I was taken to in Siberia was a huge drab, dirty and cold building about four stories high. I was given an extensive tour and at one point was led down a hallway lined with doors to many different types of research rooms. Some of the researchers were

I was given a complete tour of The Russian Academy of Science in Novosibirsk, Siberia, an area forbidden to any outsiders, especially Americans, just six months prior to the breakup of the Soviet Union. I was not impressed with the technology and lack of computerization, but the people who worked in these facilities were very well educated.

experimenting with unknown combinations of elements and had beautiful samples.

One of the scientists I met was Walter Barshai, who later created a company called Tairus based in Bangkok. They were a formidable competitor because of their low cost of operations but were very open in explaining all phases of their production. They had developed a hydrothermal process, but instead of using gold or platinum crucibles inside a pressure vessel, they used a steel pipe about four feet long. Once loaded with natural beryl crystal of low quality, plus some chromium for color, thin seed plates were hung mid length in the interior of the pipe, which was then filled with an alkali type of water then tightly sealed. The pipes were placed on commercial hot plates that ran up to 600 Celsius, sometimes three to a heater, and enclosed into a locker type metal box. The pressure in these pipes rose to thousands of pounds per square inch.

"Don't these blow up sometimes?" I asked. "Da, sometimes," was the response. I mentioned how dangerous beryllium was and that an explosion would spray the entire lab with the cancer-causing chemical. "This can kill you," I told him. "Da, yes, but that is life in Siberia," he said.

Having been to Siberia many times, I can fully empathize with his sentiment, "So what?" In December, it gets to minus 40 degrees Fahrenheit and even oil will freeze. I always made it a point of planning trips around spring or summer, never winter.

The problem with using just steel in the growth chamber was corrosion The combination of water and the alkalis at high temperature caused the steel to dissolve and become weak. It also created a strange type of inclusion to appear in the stones made by this method, gemologically called a "water in oil" effect. If a crucible is used, as in the Biron process with gold crucibles, this result is almost nonexistent. The biggest headache this process presented to me was the cost and the speed of growth, about four weeks versus our 12 months. Their production volume was much less than a flux process, however, with a few hundred carats versus Chatham's 10,000 carats per run.

We eventually opened a buying office in Moscow that cost me about $1,500 per month. We had no idea who owned the building, but figured it was the government. A guy named Boris came around once a month for cash. We didn't ask any questions but made sure we only paid once a month.

Once word got out that "Dr." Chatham was in town buying up created emerald from this new office, we had sellers coming in offering various qualities of rough emerald. All deals were in cash, and we were handing out about $100,000 per month. This emerald was not sold under the Chatham brand but to resellers all over the world. I was trying to keep it out of the U.S. market.

Unfortunately, our main source of emerald couldn't keep his mouth shut about the cash he was bringing in and some guys went to his house and beat him and his wife until he opened his safe. Forever burned. We never saw him again. We also investigated other sources we heard about from Dan Kim since we knew this one seller was not the grower, just a middleman.

At the end of the emerald wars of 1995, the Biron hydrothermal process (out of Australia) won the price battle – even beating out the Siberians – but lost the marketing advantage.

We eventually opened a buying office in Moscow that cost me about $1,500 per month. We had no idea who owned the building, but figured it was the government. A guy named Boris came around once a month for cash. We didn't ask any questions but made sure we only paid once a month.

They kept lowering prices, eventually to below cost, so it closed and was sold off. Operating from Perth, Australia, was also problematic because of the huge distances and time zones involved.

In the end, Japan pulled its Inamori Emerald out of the United States; Biron closed and was sold off; Regency and Union Carbide closed their emerald production; and I bought out the Gilson operation. The world was left with Tairus based in Bangkok and Chatham based in the U.S. Chatham walked away from all big box customers and Tairus struggled to keep up with the cutthroat-prices they helped to create. Then the big box stores walked away from the created emerald market.

It would take years to repair the damage the emerald wars caused, and a lot of the problems started right there, in Russia and Siberia.

30

Trying to Do Business in Ukraine

Previously, our plan was to get the Ukraine government to help us in this project to make diamonds, but just 10 days prior to our visit, the company was told there would be no influx of cash subsidy from the government … Their plan was to form a joint venture, with us picking up the tab. This didn't really concern me at the time, plus I felt more comfortable dealing with an individual company and the four Americans than dealing with a foreign government I knew nothing about.

What really intrigued me during my Russian explorations was the prospect of created diamond, the one gemstone my father never achieved.

One of the Ph.D.'s I met and became friends with in Siberia was Dr. Viktor Vins, who had doctorates in physics and chemistry, and whose specialty was researching diamonds. He was using the split sphere system called BARS, a Russian acronym that is untranslatable. From our many conversations, it appeared that Russia had been growing diamonds way before the U.S., but Russia being Russia, there was no fear of any legal action from General Electric or De Beers – there was no rule of law, so a lawsuit would be meaningless. Same holds true for contracts, they are worthless. I have signed several contracts promising to pay millions for diamond research, but none ever came through. My attorney even told me not to waste his time and my money to look at these contracts.

Everything I saw in Russia was for industrial applications, not gem quality. I tried working with Dr. Vins, and we grew many small poorly colored diamonds, but nothing of the commercial quality I needed. Dr. Vins did give me hope it could be done and I left him to continue his experiments and moved on.

I then hooked up with another group working in Kyiv, Ukraine, a satellite country of Russia for many years. Because of their close history, Russian was the common language until Ukraine became independent in 1991. The city was known as Kiev when I was traveling there, but the spelling changed to Kyiv, derived from Ukrainian language, in 2018. The turmoil between the two countries continues to this day.

Four Americans, one originally from Russia, were exploring the possibility of making diamonds at a facility in Kyiv already known for making super hard materials, such as cubic boron nitride and diamond dust used in abrasives. The company had the same belt type presses used by GE to make diamonds but on a smaller scale. These four people invited me to join them in their venture because they needed our capital, marketing expertise and technological help. I was interested and took multiple trips to Ukraine to see what might be possible so Chatham could develop gem quality diamonds. I insisted on seeing the operation first-hand to evaluate their expertise and technology level.

To get from San Francisco to Ukraine was typically a 15-20-hour journey, flying four hours to Chicago, eight to Frankfort, and three to Kyiv, with a few hours sprinkled in between for good measure. I normally

{163}

The living accommodations in Kyiv, Ukraine were minimal at best, however this private apartment was better than the one hotel in the area I stayed at once.

carried on everything so none of my luggage would get lost and I wouldn't have to wait for checked bags. This put me at the front of the customs line for inspection and X-ray search and usually went through customs with no problem.

One time I got through the custom process so quickly that I beat my ride by 30 minutes and was left standing in front of a sea of sharks – tough-looking, real labor men – like in Russia and Siberia who were looking at me like I was a guppy. One shouted out, "I want to take you to town, yes?" "No," I said. After about three more guys tried, I got an attitude and quit acknowledging them. Igor, one of the partners, finally showed up and rescued me.

On my first trip to Ukraine in 1994 we stayed in a 400-square-foot rented apartment that housed five people when we weren't there. They disappeared to "someplace else" when we arrived. After the hotels I experienced in Siberia, this seemed like a plus to me. The living room, with a Russian-language television, was also the dining room and a bedroom. A second room had two narrow beds: I got one and Wayne, another partner, took the other. The bathroom was closet-sized with a toilet and a tub with a hose so short you had to sit down to take a shower. There was no heat; they said it was not cold enough for the city to generate heat with steam, but it was 40 Fahrenheit outside!

A guy and his girlfriend, presumably the lease holders, showed up each day to cook breakfast and dinner for us, and then would disappear again. She prepared the food in a kitchen the size of a closet with one light bulb. The menu was potatoes, potato soup and a meat-mash fried in old grease. The pans and leftover food were not stored, but it was so cold I don't think it mattered. Everything was old, dirty and poorly constructed.

The head of diamond research from the company we are trying to hire came to the apartment at 6 p.m. our first night. I had been up for 35 hours, and the conversation was all in Russian, so it was tough to stay awake. I opened one of my precious bottles of wine to wake me up. Two hours later they left, and we reviewed what was discussed.

Previously, our plan was to get the Ukraine government to help us in our project to make diamonds but just 10 days prior to our visit, the company was told they had to earn their own keep. There would be no influx of cash to help them survive. The company's plan than became a joint venture, with us picking up the tab. This didn't really concern me at the time, plus I felt more comfortable dealing with an individual company and the four Americans than dealing with a foreign government I knew nothing about.

We were picked up at 9:30 the next morning to go to the company facility, which was about 30 minutes away. Everything was gray; the sky looked like it was ready to unload, and the wind picked up, so I was freezing in the car. I knew it wouldn't be any warmer in the factory.

We entered a huge building with no lights, heat or an elevator to save electricity. We were led into a conference room that sat about 30 people around a long oval table that looked like some kid made it in wood shop and only got a "C" grade. Light was coming in, but the windows were gray with dirt and dust.

The president of the company came in and was all smiles as he pumped my hand, "I am in the presence of a legend," he proclaimed. Previously, my "partners" had only worked with the crystal growers of the company, so had not met Mr. President before.

All the diamond news I had been generating with samples already grown elsewhere (with Dr. Vins) was reaching his desk, including my spots on the United States television shows (Dateline and CNBC) where I shared slides of his equipment before I made the trip to Ukraine – I wanted to see some proof of what they had before I got on a plane or invested time and money. He said he was surprised to see pictures of his production facilities on TV and said if anybody who knew him saw it, they would know where it was.

"That's the price of publicity," I told him. "If you don't want to become well-known, perhaps we should stop here." He ignored my comment and gave a 30-minute speech on his changed situation. Then he turned to me: "It is now time to hear your proposal, Mr. Chatham."

They love to give "speeches" in Russia and the satellite countries like Ukraine that used to be in the Soviet Union were the same. It is similar to a statement of intent. I didn't expect to say anything, so I had nothing prepared, but I was stuck, so I told him all about our past successes, our marketing prowess and leadership in the category of created gems. I also said this partnership would only be possible if they could grow the diamonds we needed. This aggravated him, and he questioned my meaning. I told him I had not seen any diamonds of over 1/2 a carat yet, so I was skeptical of their ability to grow larger diamonds, even based on the information and technology I had already shared with them.

Soon after, we went to lunch in the "executive restaurant" without Mr. President. There were scientists and one propaganda official from the company and the four of us, all ticked off at each other because of my speech. Vodka was poured, but we didn't touch it. I had a plate of cut up tomatoes (I detest tomatoes), some greasy soup with potatoes and some very tough beef. We kicked around some ideas and proposals. How much will it cost per carat? What is the production rate potential? We were told it would all be discussed the next day.

We headed back to our apartment where someone had opened the window in my bedroom, so it was 40 Fahrenheit inside too. We opened another of my bottles of wine and ate my snacks and nuts. I did not recognize dinner.

The next morning, we went back to the company and met with the head research scientist and their marketing person and the director of sales.

This room is full of moderate-sized belt presses used to make yellow diamond powder. Our goal was to use this facility to grow larger white diamonds by enlarging the press size to grow bigger stones. It was partially successful.

"We can produce 100 crystals per month per press using Mr. Chatham's information," they told us. "We need $53,000 to start up three presses. The cost breakdown is about $45 per carat, and we will supply approximately 1,200 carats of stones already produced for this investment!"

I was overjoyed and said "OK, let's get on with it."

But first all the papers had to be drawn up to address the "protocol" of the deal. They loved that word. I already knew from my attorney that any contract I signed in Russia, Siberia or the Ukraine was worthless, so I don't care what they write up.

Then we went to lunch again: same room, same vodka, same tomatoes. This time everybody was happy, so we toasted with the vodka. I could tell the marketing guy loved this part of the job. I don't like good vodka in the United States, and this local stuff tasted like gasoline. They kept coming up with new things to toast, so I quit.

After lunch, we went back to the main building through a park-like setting that had seen much better days. The fountains were all broken and empty; the flower beds overrun with weeds; and the building looked dilapidated with broken windows, missing tiles, marble stairs that were worn down through years of use. Marble is not a good material to use for steps because it quite soft compared to granite, but they used what they had.

We went to another meeting room to discuss how much the company must make on this

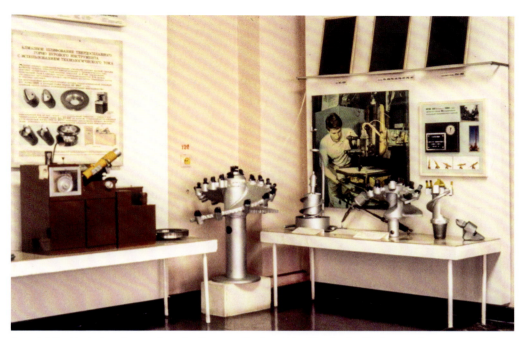

An exhibit of industry diamond applications on display at the Ukraine facility I was trying to work with. Most people don't know just how important diamond is to machining (finishing) any metal. Before the mid-1950s, all industrial diamonds came from low-end natural diamond, but due to the low quality of the crystals, which were often black and crumbly, their performance was poor. Lab-grown diamonds soon pushed natural diamond out of the process and all machining, saw cutting or anything that requires a hard surface to finish uses lab-grown diamonds. It is about a 5 billion carat per year business.

venture. Then we argued for five hours over why we couldn't pay more if this was to succeed at all. They wanted twice the rate we agreed to this morning. I got up and said, "Thanks for lunch, I do not need this project, you do. Let's get out of here."

They threatened to accept a deal with H. Stern in Brazil, and I told them, "Go ahead. I know Stern personally and he wouldn't touch this with a 10-foot pole!" I knew this was true, having met H. Stern and Amsterdam Sauer at a talk I gave in Rio De Janeiro many years prior.

This "deal" was making all my alarm bells go off. So far, I was covering all the travel and lodging expenses, plus providing my technological expertise, but we had not arrived at a cost for the diamonds or if they could even produce what we needed with their equipment.

The next day was Saturday, so no company visits. Breakfast was some type of pancake stuffed with cottage cheese – yuck. We walked into town and went into a jewelry store of sorts, with a bunch of flame fusion rubies set in brass or silver. I took out my camera and moved closer to the exit and you guessed it, the guard had a fit. Igor had to get me out of trouble with the store security, like he was guarding the Czar's golden eggs or something. I think I had enough cash in my pocket to buy out the store.

{167}

I was unimpressed with Russian technology, such as this panel of old-style (1940s) temperature controllers in Kyiv. How did they get a space ship to the moon I wondered when I saw this.

We got out of there and walked around the shops. More stores and street vendors existed then on previous visits, but it was all junk. Cheap vodka was sold everywhere for next to nothing. For lunch, we each had a beer and five cans of peanuts. Back at the apartment we found a chocolate cake on the kitchen table. "What's the cake for?" I asked. "Is it someone's birthday?" No, it was dinner. They eat whatever is available in the stores. Keep in mind this was in the early 1990s and "retail" was an emerging concept. There were no supermarkets or even small grocery stores, yet.

We traveled to the airport the next day in an old Lada, the only car made in the Ukraine. It came in one model and one color. We passed through a few check points, and I asked Igor if I should wave at them. "No ... don't do that!" he said. They pulled over the car behind us with a wave of his machine gun ... so I waved!

At the Ukrainian airport, with one agent and hundreds of passengers, I declared that I was holding $1,065 in U.S. dollars, so the guy asked to see it and counted it. "Only $1,064

... what are you trying to do, smuggle?" he asked me.

"Just a mistake," I mumbled. I went through the X-ray machine, a search of my bag, and then I was at the gate. A security guard asked for my ticket and passport, which he handed to another guard and told me to follow him. I asked for my passport back, but he said "No, move now."

The guard walked me down four flights of stairs into the bowels of the airport where I meet four hung over KGB-like officers in long black leather trench coats. They looked just like the Gestapo in an old B movie. They pointed to a bag, "Is this your bag?" I said yes, and they opened it and pulled out a can of cleaning fluid. This is carbon tetrachloride, commonly used to remove spots on clothing. The guy poured some on the ground and after some effort, got it to burn. I know where they are going with this and I said, "I'm sorry, I didn't know it was in there, throw it away."

"No, you must pay fine in Rubles and sign protocol," they told me.

I told them I didn't have any Rubles, only U.S. dollars.

"No good, you must go to the bank and change to Rubles."

"I can't go to a bank. I have no ride. I don't speak Russian," I said. "Can't we work this out some way?" This is how you bribe someone without making it look like a bribe.

"You must sign protocol," they told me and filled out a long form that might have said I killed 83 people and agreed to the gas chamber, or a firing squad.

"Do you have scripts?" they asked.

I am not sure what a "script" is, but figured it must mean money, so I said, "No, just dollars," and pulled out a $50, my smallest bill, unfortunately.

"The fine is $1,250,000 Ukrainian dollars," they said.

I have no idea how much that is. "How much is that in dollars?"

"$10 dollars," they said, so I showed him my $50.

"You must go to town to change," they said.

"No, I will miss my plane. I have $50, take this, and have a party, forget the change. OK? I leave now?" and started to flap my arms like a bird. "I want to go home."

They shuffled around a little and said, "We can't give you a receipt for this...."

"No problem," I declared.

"OK, take him back up and give him his passport and ticket back, we shall forget about this incident," one of the guards declared.

I had just been "shook down" Ukrainian style. When I got to Frankfort and in an airport hotel, I thought I had died and gone to heaven. And guess what? The cleaning fluid was still in my suitcase! I tossed it.

So far, I had sunk about $150,000 in expenses into the Ukrainian diamond effort and had little to show for it except excuses and broken promises. Some funds went into the factory for parts, but not much because I could not see any results. I felt like I was being used, big time.

This was one of many trips I made to Kyiv over a two year-period.

31

Introducing the Chatham Created Diamond

First, Richard T. Liddicoat, president of GIA and an old friend, came up to the booth. "Tom, you can't do this. You shouldn't do this, it's not right," he said. "Richard, this is what I do, I grow crystals. It is the holiest holy grail for any crystal grower!" I told him. He walked away shaking his head.

It is May of 1996, just before the JCK show in Las Vegas and I go to Ukraine to try again to make something happen with diamonds. I take off from SFO at 8 a.m. on United business class because I have so many miles, I get bumped up a lot. Then, from Washington, D.C. to Frankfort on Lufthansa, then into Ukraine, also on Lufthansa. Igor joins me in Frankfort. As usual, the foreign airlines care little of my flight status and declared my carry-on luggage in "excess" at 16 kilos. "Only 8 kilos allowed. All you Americans pack way too much!"

Eight kilos is about 18 pounds, which is not enough for my computer, some food and a bottle of wine. So, I had to go back to the ticket counter, retrieve my baggage and put the extra weight in my suitcase. Since it all goes on the same plane this makes zero sense to me, but I had no choice.

We make it to Kyiv and the airport looks the same; the vultures outside look familiar. This time it's hot, very hot and humid. And everybody smells. Our ride shows up, an old Lada that I swear could not run. Remembering how dangerous they drive in Russia and Ukraine, I pass on the front seat and climb in back. So, we ramble off, coughing and sputtering, cachunk, cachunk, on every pothole because the shocks are shot or non-existent.

On the way out of the airport, we go through a checkpoint, and I salute the guard as we pass, I thought Igor was going to have a fit. "Don't do that" and turns around to see the cop waving his baton at us. We don't stop. There is no air conditioning, of course, and none at the same apartment we rented last in our previous trips. Same room, same sheets, I think. It looks like I left yesterday. My trips to Ukraine are spaced about four months apart with infusions of cash from me to the factory plus travel expenses for those who traveled with me. The trips are to evaluate progress and see what has been produced.

The first thing the group discusses is the meeting they had yesterday at the factory. Everybody is concerned I will quit because little progress has been made in either quality or size of the diamond crystals produced so far. Then they pulled out about 25 small bottles, each containing 3-6 stones. All are white, some are opaque, some very included, but all are white. Igor says, "You must be pretty disappointed, yes? There are so few stones for six months work."

{171}

After a few minutes I answered, "I am disappointed in the lack of information and communication. That almost sank the whole project. But I am very pleased with the first results. We now have something to sell!" No one in the world to our knowledge had made any white diamonds so far, on a commercial basis, so this was "one small step" toward our goal. We open a bottle of wine and celebrate.

First thing in the morning, someone comes in to make breakfast. We had mashed potatoes, tomato and cucumber salad, fried veal and cheese pancakes with raspberry jam. I can't figure out what you're supposed to eat and at what time, so I just eat what I can and don't ask. I also know what to expect at the factory later, vodka and tomatoes.

One big problem on this trip is that there is no hot water in the building. "Somebody didn't pay the bill," our host tells me. The first day, I just wet my head down – there's no way I'm hosing down with freezing water. I don't even think about how they wash dishes, probably just polished the grease! By this trip, I have learned to pack lots of snacks and a few cans of tuna, chicken, nuts and wine.

Our driver lets us know he is "delayed" at the gas station because of a long line and will be there soon. He shows up three hours later. The sun is out, everything is green, and everybody smells. We get to the factory and stop at a huge gate: The guard wants to know who we are and why we are here. I tell them to say we are looking for real estate for a new disco. They tell him something and I salute when we pass him. The guard looks like he has been on a bender for weeks.

Mr. President and the official protocol man are in Moscow today, so we meet with the head technician. We assemble in a room that must be 100 degrees Fahrenheit with 90 percent humidity. There is no air conditioning in

this building, and we are all wet in minutes.

Every meeting is started with a "protocol" speech by the leader of the meeting, followed by the senior visitor, me. All must be translated. The speech is given in a third-party way, even though the speaker is talking about himself. I hear all about the problems with the cold winter, the freezing pipes, the lack of funds to complete the work required, etc., etc. "So now, Mr. Tom, what is your statement." This guy understands English, but always uses a translator, a common practice. It gives them more time to think.

Having gone through required protocol speech several times already, I give 10 minutes on my unhappiness at the initial results according to the contract I signed on a previous visit, but I am encouraged based on the evidence so far. We look forward to improvements in quality and quantity to make this successful. I feel like I am talking into an answering machine but six people are watching you do it.

We are joined by the PR man for lunch, which means the vodka is about to flow. We go to the same lunchroom as before and see the same food: sliced tomatoes on cucumbers, borscht soup with a layer of fat floating on top, deep fried breaded beef with greasy potatoes. I think I have about 600 grams of fat in front of me, then the vodka. Everybody is happy, so there is one salute after another. Then the PR guy outlines the schedule for tomorrow:

We will pick you up at 11 a.m. and have a meeting. Then lunch. Then we will discuss the contract and then Tom will go with the head technician to see the progress we ordered on the presses and then we will meet with the president.

Our "car" arrives, a van from the factory.

Another security check by the guard, a hungover KGB-type cop. "Why are you back?" he asks me. "I have come to steal more of your excellent toilet paper, it is so different, like parchment." He can't understand a word I'm saying, but he can tell by the howl of laughter that he is the butt of a joke but he waves us through.

The next day, we go into the same sweat box office and have hot tea and vodka. Today we have two new people, an official translator and the economic officer representative from Kyiv. We immediately get into the area I was dreading.

"So, how much are the diamonds worth?" they ask. "Our advisors say they should be sold at half the price of natural diamond."

I had the feeling the contract was about to be re-negotiated, with this first parcel to start. I have learned this is a common occurrence with Russians, renegotiating after a deal is made, which I find very annoying.

"These crystals just produced are almost uncuttable, curiosities at best," I tell him. "If you want parity with natural, I would say $5 per carat." We need more pressure and temperature control to produce anything of greater value. People are not following my instructions correctly."

This threw them into a tailspin, and they jabbered for 15 minutes and then suggested we grade the rough based on quality and price them. Meanwhile, security has come to take me into the labs to inspect their work. I have my camera with me, but after all my experiences trying to record anything even slightly sensitive, I leave it in my bag. I did manage to sneak off a few shots.

We go into the factory and there are three presses marked No. 1, No. 2, No. 3. I am told next year they will stencil "CHATHAM" on them. I was given a complete explanation of how the press is set up, with all the components laid out for growth in a chamber of about 2 inches square. Of course, I am very familiar with the press set up, already. What I am really looking for is any signs of improvements my money was to have been used for and I see nothing new. So, after about 30 minutes we go back to the meeting and I comment that I see no changes to the set-up, no new equipment.

"You must be blind, Mr. Chatham, everything is different," I am told. I let it go, everything looks old in Ukraine. What I did see quite clearly was that this "laboratory" was from the stone ages. Nothing digital, no computers, all 1920s electronics and engineering. They had no reliable way to control high temperatures at high pressures. This was not the new equipment I was looking for.

The contract is pushed over to me with a written document: "We must have protocol statement from Mr. Chatham":

1. What is your opinion of the work completed?

2. What are your plans for our future?

3. Please give us a summation.

We talked for about 30 minutes, and they stated they would be up to 200 pieces per months in three months. When I asked if he would put that in writing, he didn't answer. The issue of grading the stones was avoided, even though one of us had separated the crystals into those that could be cut, those that could be sold to collectors and those that were junk. All the crystals have a fair amount of iron crystals growing on their surface. These are just collector's editions at best and I am already thinking about how to create

{173}

a certificate for each with a number, the date grown, and how to mount them on a plaque of some sort. There are about 100 crystals, some up to 1.50 carats.

Now the president of the company arrives. Last time we saw each other, he left in a huff. This time I was prepared to tell him I did my part of the deal, but he screwed up his. He comes in all smiles, gives me a big bear hug and completely disarms me for battle. He orders out the cognac, not the cheap vodka and we have small talk for 10 minutes. He has read of all my exploits and articles. "Everybody in Europe is talking about the great Tom Chatham and his overthrow of De Beers!" He did not explain what exploits he was referring to but it must have been the television interviews I gave. Then he gets serious.

"It is time for your report, gentleman, begin."

The PR guy introduces the contract man who introduces the head of research. He gives a 15-minute talk, in Russian of course, and then I am asked for my state of the union speech. I give them the same as before, which is a good thing, because the contract man produces a written version they collected from my earlier visit and typed it out, then mimeographed it! Yes, mimeographed, not photocopied.

I get the feeling that this is important, so I get my green pen (it was a Chatham habit to use green pens and green type in letters, which I learned from Ed Coyne) and sign the three documents, then so does the president and a few others. Feeling like Clinton, I hand my 59-cent green felt pen to the president; he in turn, hands over his pen. I have just traded a 59 cent pen for a $5 pen and feel great! More hugs and good-byes and we leave.

Back at the apartment, we open some wine, my last bottle, and look at the crystals. They are not very good but show they have been

trying. Dinner is potatoes and chicken "something."

The next morning, we have sardines, pickles and egg followed by a dozen smoked fish, finished with chocolate cake! These people eat the weirdest things in unusual combinations. Today I am leaving Kyiv and hope I never have to return. The partners can oversee it on their own if they need to.

I have decided to smuggle the diamonds out of Ukraine in order to have them in time for the JCK show in Las Vegas. The factory was very concerned for my safety and gave me a phony declaration to use in case of a severe emergency. "Don't use this unless the cuffs are about to go on," they warned me. Moving diamonds requires a permit in Russia and Ukraine and to be caught with them would be very serious trouble for me.

So, I go through customs check point No. 1 and the official asks, "what was the proposed business discussion you were in?"

"Scientific," I answered.

"What scientific," he demands.

So, I go for broke: "We discussed the epitaxial deposition growth of bucky balls on non-carbon surfaces under less than super high pressure at room temperature." He had no idea what I was talking about and neither did I, but he took it hook, line and sinker.

Stamp…Stamp.

The other check points are no problem except they tore my computer and camera apart. I got back at them by stealing power for my laptop in the waiting area. All the guards kept looking at me wondering what law I was breaking. Nobody had a laptop over here or a camera with five lenses. On to Germany, food and warm showers!

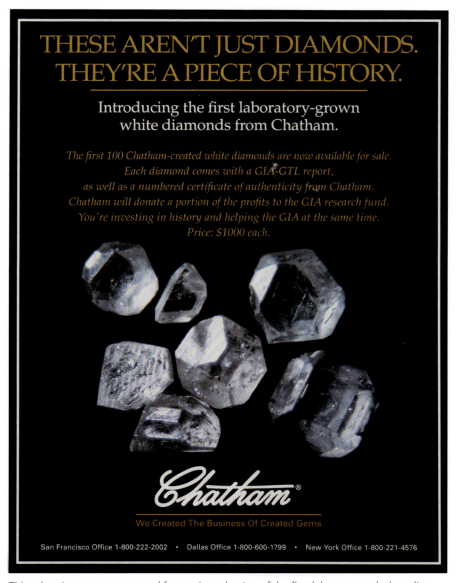

This advertisement was created for our introduction of the first lab-grown colorless diamonds at the 1996 JCK show in Las Vegas. It ran in many of the industry jewelry magazines and was the first lab-grown colorless diamond introduced into the jewelry trade by any company. This has evolved into a billion-dollar industry worldwide.

Faceting a laboratory-grown diamond is no different than cutting a natural diamond. At Chatham, however, we strictly follow the rules for producing the most attractive cut stones at the cost of lower yields. Poorly cut natural stones are common and command lower prices and often must be re-cut. Chatham strives for what the industry describes as "Ideal" cut where all of the facets are cut at exact angles for maximum brilliance.

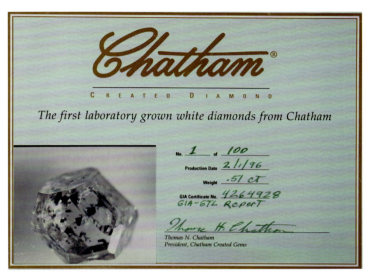

Chatham's first Lab-Grown Diamond certification.

When I return to the U.S. the JCK show is weeks away. I sent off about 50 crystals to GIA for certification and had our ad agency put together a certificate that I numbered plus a picture of the single diamond rough with the cert attached to sell at our booth at the JCK show in Las Vegas. We had a press conference where I announced that Chatham was going into the created diamond business with white, colorless Chatham Created Diamond and would be selling our first samples of production at the booth.

What an uproar. First, Richard T. Liddicoat, president of GIA comes up to the booth. He is an old friend; "Tom, you can't do this. You shouldn't do this, it's not right," he said.

"Richard, this is what I do, I grow crystals. It is the holiest holy grail for any crystal grower!" I tell him.

He walks away shaking his head.

Then, the head of the Diamond Promotion Services, Preston Foy, comes up, looking like a deer in the headlights; "Tom, what does this mean? What are you doing?"

"Well for one, you now have job security Preston; you will need to advertise natural diamond all the more," I say. He buys one of my samples for $1,000, the asking price no matter the size. I am sure it was in De Beers hands within 48 hours.

Then Martin Rapaport, chairman of The Rapaport Group, which provides services to the natural diamond and jewelry markets, comes over to the booth, steaming. "You can be stopped, Chatham, you can't say these are diamonds!"

"Martin, want to read the GIA report on one?" I ask. He marched away all flustered. I can't tell if he is mad because he thinks I am lying or just mad because I accomplished it. You can never tell with Martin Rapaport.

Thus began the long and winding road to what is now a thriving billion-dollar market, but not all Chatham grown.

32

The Partnership of Tracy Hall and Chatham

This was a huge jump for us working with the Halls in Provo. And with Tracy Hall's reputation behind us, we had a sure winner on our hands. Nobody trusted the people in Kyiv, and I felt ripped off. I never could see where the money was going and was often told of equipment installed but I never saw it.

The year between 1996 and 1997 was hectic with all the talk about our created diamonds. I was interviewed on "Dateline" on NBC, CNBC, ABC, a few radio stations and in many jewelry magazine articles.

Out of the blue, I got a call from David Hall Jr. in Provo, Utah. His father, Tracy Hall was part of the team at GE in the 1950s that first discovered the process for making diamonds in what they called a belt press. David suggested we team up to make diamonds in the U.S. I thought the idea was priceless: Two legends working together would be a fabulous marketing story.

I flew to Provo and met with Tracy Sr., David and Tracy Jr., who all had Ph.D.'s in chemistry or physics. My visit with Tracy Hall Sr. was remarkable. I showed him what we grew, the best of course: A 1,000-carat ruby and a nearly 500-carat emerald and some of our white diamond rough. He was amazed, especially at the growth time it took. "Did your father or you make any money doing this work?" he asked.

"Yes, he did, and we continue to make profit from these products," I told him.

"I got nothing from GE in 1954 except a $25 savings bond. So I quit," he grumbled. His wife spoke up, "Now dear, they took good care of us, and you did work for them, so that was the rule."

I told him of my father's run in with companies and even his alma matter, Cal Tech, trying to claim ownership, but my father had them all shut down when he could prove prior knowledge and experimentation. Tracy quit GE and went on to create a high-pressure company in Provo to make diamonds as well as becoming a professor at Brigham Young University. He showed me his original press and a few second- and third-generation presses he kept in his office. All were idle, but were definitely a piece of history. I was able to get one donated to GIA for their display in Carlsbad. It was a very generous gift.

Next, they showed me their own press design and how they were using it to make a super-hard, carbon-like material, called carbon nitride, for oil drilling bits, some bits 12 inches in diameter holding hundreds of these "teeth." They suggested that we bring over our Ukrainian crystal grower and set him up to make white diamonds. They proposed a position with salary, a place to live, and two Russian-speaking undergraduates from Brigham Young University to help translate.

Real Diamonds Made By Men in GE Lab

LABORATORY DIAMOND MADE BY GENERAL ELECTRIC
. . . Dr. H. Tracy Hall, Schenectady GE chemist, displays manufactured diamond.

Tracy Hall had a Ph.D. in physics and worked with a team of scientists at General Electric to figure out how to grow diamonds in a laboratory. He invented many presses, most notably the "Belt Press" that stood about 20 feet high. Although it was a joint effort, Tracy's experiment produced success in 1954. Feeling unrewarded for his efforts, he quit and joined the teaching staff of Brigham Young University in Provo, Utah, where he continued his diamond research. General Electric was frustrated that Hall left, even though the patent rights where in its name, and had some of his presses confiscated. Credit: *Press and Sun-Bulletin*, Feb. 15, 1955.

This six-sided press was invented by the Tracy Hall family in Provo, Utah. Each of the huge computer-controlled hydraulic cylinders pressed inward against an anvil containing graphite and other elements. Because of its unique six-sided design, no one cylinder had to do all the work to achieve pressure nearing 1 million pounds per square inch. Diamond and other super hard materials would form at temperatures exceeding 1,200 Celsius. Each press weighed many tons and was 12 feet high.

These images show views of the Hall press from several different angles.

This was a huge jump for us working with the Halls in Provo. And with Tracy Hall's reputation behind us, we had a sure winner on our hands. Nobody trusted the people in Kyiv and I felt ripped off. I never could see where the money was going and was often told of equipment installed but I never saw it. Moreover, every lab I went in was archaic by our standards because everything was mechanical; no electronics or computers were used. I was shocked when I visited a lab in Siberia and saw someone using an abacus for a math problem. No calculators were in existence in Russia or Ukraine. Russia and Ukraine were unsophisticated when it came to engineering; computers were still out of their reach and electronic controls were very sloppy. This marriage with the Halls was made in heaven.

Diamond growth requires extremely high pressure nearing one million pounds per square inch, at a temperature of 1300 Celsius, so each affects the other. Without computers, you're bouncing all over the place, which is not good for growing crystals. No two runs were ever the same. The set up in the Halls' laboratory was completely computerized and able to control temperature and pressure to a very finite degree.

Sergey, our crystal grower in Kyiv jumped on the huge opportunity to work with the pioneer of growing diamonds and the chance to use the best equipment available. I had little trouble, with his education and specialties, in getting Sergey a work permit to enter the U.S.

We flew him over and he stayed at my home in Hillsborough, California. I built the house in 1992 and at about 6,000 sq ft, it was slightly overwhelming to Sergey. Going into a Safeway store was like walking through Fort Knox for him. He had never seen such an abundance of food or lived in such a luxurious setting. I eventually got him to Provo and moved into a small house.

He began his research with the Hall brothers with their six-sided press. I hoped that what we accomplished in Kyiv combined with the knowledge possessed by the Hall family would produce significant improvements in diamond growth. Unfortunately, 30 days into Sergey's move to Provo, I get a call.

"Tom, we have a problem. Sergey wants to go home for a visit, he is homesick," David Hall told me. Homesick, I think to myself? What the hell could he be missing in Kyiv? I jump on a plane and go to Provo. I found Sergey at the lab. "Sergey, what are you doing? The U.S. does not have a revolving door for people from your part of the world. If you leave you might not get back in." But he leaves anyway, saying he will be back in a week.

Two weeks later, I got a call from the U.S. State Department checking on Sergey's application to come back to the United States. I explained his circumstances: "We are 100% responsible for his support. He is very well-educated; he is a needed immigrant." The State Department was not impressed. "We have heard all that before, the entry permit is denied," they told me.

The whole project collapsed. Sergey never got back into the United States, and we found out later he was missing his girlfriend in Kyiv. Not a wise choice, in my opinion, but then again, I don't know how he was doing in the Halls' lab either. There is so much we take for granted in America versus an undeveloped country like the Ukraine, perhaps he just couldn't fit in. All the new technology must have overwhelmed him and working with equipment he had never seen probably didn't help. He went back to his old job in Kyiv where he felt comfortable and in charge, but never did achieve much in diamond growth after that.

This experience, both with Sergey and the Halls taught me a lesson about success: You have to want it, have a passion for it, and be willing to work years to achieve it. This assumes you have the talent, as in high-pressure chemistry, and assumes you have the facilities to achieve it. The Halls had all the above but lacked the passion and patience it would take to grow large single crystal diamonds suitable for the jewelry trade. It is a huge step from making yellow diamond powder to growing multiple colorless crystals in multi carat sizes.

My passion was for gemstones and the jewelry business. The Halls were good at producing drill bits, not gemstones for jewelry. Sergey was the same.

{183}

33

The Rebirth of the Japan Connection

"Mr. Chatham, we are prepared to offer you the opportunity to work with our technicians but under one condition: No one must know the source of the diamonds."

— *Mr. Kuzomoto*

Fast forward to 1998. I had the diamond crystals from Kyiv and had gotten a lot of recognition and exposure from them, but I was through with Ukraine. I did not invest a huge sum into Kyiv. I thought they were devious; their equipment was low-quality, and I could not work closely with them. Added to that, the president of the firm was not too happy with me after we hijacked Sergey to Provo, Utah. The start up with the original four Americans collapsed when I cut off funding. They reluctantly agreed with my conclusions and summations of our progress in Kyiv.

Miraculously, I received a phone call from Mr. Kuzomoto in Japan about this time.

"Mr. Chatham, do you remember me," he asked. It had been almost 10 years since I met with him in Tokyo, Japan.

"We have changed our minds and want to join you in the diamond business, are you still interested? Can you come to Tokyo?" he asked.

I couldn't believe it; they came back, and the timing couldn't have been better. The demand was already there, and we were losing momentum.

It was not so formal with Mr. Kuzomoto or in the board room this time. I met with the engineers responsible for diamond research, a president of the division and Mr. Kuzomoto, the director.

My first question was, of course, "Why the change of heart?" Without going into too much detail he said, "Our diamond competitors did not keep their word and cut our legs off in a product category that cost us $50 million in sales."

"I am not surprised, Mr. Kuzomoto, they have been ruthless for over 100 years and cannot be trusted," I said. "Their reputation worldwide in the trade is one of great caution. They are cold-blooded, to say the least."

"Mr. Chatham, we are prepared to offer you the opportunity to work with our technicians but under one condition: No one must know the source of the diamonds," Mr. Kuzomoto declared.

That really surprised me. "How can you be afraid of a company so much smaller than yours?" I asked.

"We don't want another expensive episode to happen unnecessarily, but also feel the need to be careful," he said.

{185}

These were some of the first rough diamonds produced in Japan. The quality was much better than efforts in Kyiv, but still fell far short of a predictable process that we could count on to reproduce.

This was fine with me. "So, we are calling all the stones Chatham Created, correct?" I clarified.

He answered "Yes," and I thanked him for his willingness to work with me and trust us to do a good job. That was the last time I saw Mr. Kuzomoto. From then on, we met only with engineers and executives, usually around 10 of them, all smoking up a storm in the meeting room. I had quit years prior, so this was always debilitating.

All the stones were made using the belt press invented by Tracy Hall for his High Pressure High Temperature (HPHT) work at GE, which stood about 20 feet high. I shared everything I learned in Russia and Ukraine with their technical staff, and they eagerly enjoyed this expansion of their diamond knowledge. Growing small industrial diamond is totally different than growing large gem diamonds.

I asked the Hong Kong cutting factory for help in finding diamond cutters. Colored stone cutters never cut diamonds and diamond cutters never cut colored stones. The two disciplines just don't mix due to hardness differences in the gemstones. Diamond is about 100 times harder than its closest gemstone brother, ruby, and only diamond can cut and polish diamonds, which requires an understanding of graining and direction of hardness. Therefore, the training to cut diamonds is entirely different than learning to facet any colored gemstone.

We located a cutting factory in Guangzhou, China, an old city of 10 million plus that used to be called Canton by the British. Cutting costs were $25 per rough carat, quite

> "One thing I really learned from my father was taking chances and making changes. If you set up multiple runs and get the same results time after time, you're going down the wrong path. You have to take chances to learn, and the Japanese refused to change their processes."
>
> *- Tom Chatham*

different than colored stones, which charges by finished stone weight. This was not as inexpensive as India but was more controllable in Hong Kong.

We worked out a price list based on size and color: yellow, blue and pink. I wanted to grow crystals that would produce at least a one carat of finished stone, but this size minimum was rarely met. We found that the rough being grown was full of metal inclusions and was difficult to cut. Our yields were low, about 25% or less, which really cut into profits. There was also a problem with inclusions and color variations. The yellows were consistently canary yellow, but the blues and pinks were all over the map. Some of the blues were almost black while some of the pinks looked more like salmon. Most of the stones were grading out as I1 (GIA terminology for stones heavily included "EYE-ONE") – a very low grade in the diamond world. We were not making much more progress in size then we experienced in Kyiv. Somebody was not following my instructions!

If it wasn't for the novelty of offering some of the first created diamond on the market, we would have gone broke if we had continued. Many millions of dollars were spent chasing this dream.

Every shipment of rough came to San Francisco first so we could evaluate the progress in crystal growth. Every shipment was the same. Progress in colorless diamond was not being made. The head engineer was especially stubborn and insisted they were all acceptable and very good. I made many trips to Japan to discuss the poor qualities and the cutting results. This was costing us millions of dollars. Why was the growth not getting better? After a few years I reached my breaking point and almost slugged the engineer who said the goods were fine. My comments were usually the same, "You have made zero progress in refining the process and I am nearing my breaking point."

The sizes of the rough they were sending us rarely allowed us to cut a carat size stone and this is where we felt the market's future lay: bigger, cleaner stones at a cheaper price than natural diamond. Cutting half carat stones in I grade (very included) was not in my marketing plan.

One thing I really learned from my father was taking chances and making changes. If you set up multiple runs and get the same results time after time, you're going down the wrong path. You have to take chances to learn, and the Japanese refused to change their processes. I finally concluded it was because of their ancient culture; the Japanese are very structured and disciplined, and if they have the correct formula, they can beat the world in perfection and quality. But if they have the wrong recipe, no one will stick their neck out to try something different. You will be blamed for bad results if you make changes, but not if you stick to the old recipe. Basically, they were ignoring all my theories and instructions.

This was a painful and expensive learning process for me. Luckily, the colored diamonds were selling well, but starting to slow down, especially in yellows. We also had some competition from a company in Florida named Gemesis. They picked up a dozen BARS pressure units in Russia after a

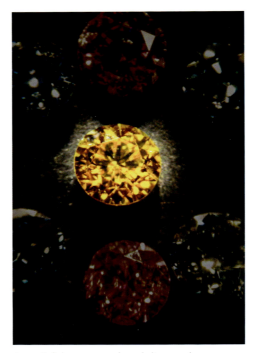

Japan's lab-grown colored diamonds were full of metal inclusions and were difficult to cut. Our yields were low, about 25% or less, which really cut into profits. There was also a problem with color variations.

The yellow lab-grown diamonds were heavily included. They were photographed while being examined at GIA. From left: 0.53 carats and 0.11 carats. Credit: Robert Weldon/GIA.

deal I had put together fell through. Overnight the Russian sellers jacked the price up from $30,000 for each press to $300,000 and I walked out. I was glad I lost it because Gemesis found out it was junk engineering and they had to make all new presses. They never did get production and quality off the ground and eventually closed, selling all their presses, 200 at last count, to 2A Technologies in Singapore who needed presses to color treat their Chemical Vapor Deposition (CVD) grown diamonds. Last I heard, Gemesis had burned through $30 million in seed money and sold out for pennies on the dollar. Again, I felt that research was short changed and misguided. Instead of building hundreds of presses, money would have been better spent researching the growth processes.

The Japanese contract was reviewed every year, but after 4-5 years and millions of dollars spent, I said no more. I refused to continue unless there was proven progress in crystal growth. We were not making progress and running out of money. Customers were not buying yellows as before and had become very selective in what color blue or pink they found acceptable: cornflower blue and bubble gum pink was the expected norm. Approximately one out of 30 stones met these criteria. I told Japan they either had to change their approach to crystal growth or I would pull out. They, in turns said: no cash flow, no more experiments.

I had an agent in Tokyo who was instructed to keep tabs on the company by taking out the engineer for sushi once a month. Then, I would be given a report: "Since Mr. Chatham will not fund research, we have no new stones, and all work has stopped." They never budged and neither did I. They quit the gem diamond business because no one would take any chances and change the processes as I had instructed. A few years later, others be-

gan evolving and producing not just colors, but beautiful white stones up to 10 carats cut, D color and VV^S quality. I sent my old Japanese partners a clipping with one of the reports emerging from Russia and said, "This is how you make progress in research; you must experiment and take chances." They never answered.

Meanwhile, companies were sprouting up all over the world making diamonds. We have since invested both in CVD and HPHT processes in joint ventures in China, Russia and India.

Diamonds were becoming a commodity worldwide and nobody had a system set up like GIA did for grading diamonds. It soon became a very necessary part of any diamond deal for stones to have a GIA certification. GIA went from a small educational facility to a powerhouse paper mill racking in over $100 million a year. At first, GIA refused to grade any laboratory-grown diamonds, as did many other labs. Because of our training at GIA, however, our company was able to use its grading standards. Eventually, GIA and others began to grade lab-grown diamonds. It really didn't matter to us, since we developed and use our own grading reports today.

34

The Evolution of Chatham Created Gems & Diamonds

The accomplishments of Chatham Inc. are extensive, starting with the first commercial flux emerald in 1938 and the first flux ruby in 1958. Carroll Chatham was a pioneer in developing the first flux gem in many types, and Tom Chatham and his brother John also made many contributions along the way.

The evolution of Chatham gem growth: some were by accident! Several of our crystals were developed in just a few years while others, like ruby, took 10 years to figure out.

Ruby (1958)

My father was struggling with ruby growth when I joined him in 1965, but his ruby research came about years earlier because of the ¼ inch alumina powder that lined our crucibles. Every time my father experienced a flux leak during an emerald run, the alumina always turned pink. He finally concluded that the flux with chromium content was reacting with the alumina, the basic chemical makeup of ruby!

My father was far from producing good quality flux ruby in 1958, but it was the beginning. Slight changes were made to the flux make up and of course, no beryllium oxide was present. Low quality natural ruby was very common and was used for "feed stock" while seed material was derived from flame-fusion ruby.

Gemologists continue to argue with me over the statement that a flame-fusion synthetic ruby is identical to the flux-grown ruby. Nothing could be further from the truth. Because of the growth method of flame fusion, the crystal lattice is a huge corkscrew and not a true crystal as it is in nature. Natural crystals do not grow in circles!

This was very evident to us when we tried to grow ruby from flame-fusion seeds: all growth was completely disorganized and haphazard and did not reflect the true crystal system of natural ruby. These hurdles were overcome using other sources of seed material and resulted in one record ruby crystal weighing in at more than 3,000 carats! We kept that for our office museum.

Crystals this size are impractical because they won't fit in the diamond saw and each saw cut cost us money in lost material, so 500 to 1,000 carats is our target.

Alexandrite (1972)

Alexandrite is a very rare gemstone originally mined from the Ural Mountains of Russia and named after Alexander the Great. Additional deposits have been found in Brazil and Sri Lanka. It is a unique color change stone in the beryl family named chrysoberyl.

We were working on alexandrite in the early 1970s, given its closeness to emerald, which is also in the beryl family. We announced our success in growing alexandrite to the trade in 1972 and much to our surprise, a company 20 miles away from us, across the bay from San Francisco, simultaneously made the same announcement!

Dave Patterson called his company Creative Crystals and I immediately called him up and introduced myself. He invited me over to his laboratory and I told him I could not reciprocate, no one goes in our labs, so he didn't need to do so. But he insisted and I, of course, immediately took him up on the invitation.

We met and he showed me his facilities, all hand fabricated like Chatham's and all up to date as far as I could tell. He was also using a safety box for handling the beryllium powder required to make alexandrite. Free state beryllium is highly toxic and must be handled with great care.

I could see this was a sophisticated operation and would be a formidable competitor. But the biggest hurdle was not his alexandrite, but rather the marketing of it! Unlike emerald, ruby and sapphire, most people in the 1970s had never seen or heard of an alexandrite, it was so rare. It is also the birthstone of June, but somehow cultured pearl took that over.

So I said, "Dave, I can see you are fully operational and making a good product. Since this will be a first rollout of a lab grown alexandrite, I am going to let you take the lead and hold off our introduction." Dave immediately understood what I was doing. "So, you're going to let me develop the market, spend a ton of money on advertising, and then Chatham will come in and scoop up the profits! Am I correct?"

"Yes, Dave, this is going to be a tough sell and neither one of us might make it work but I am giving you a free year to do so," I

{192}

The accomplishments of Chatham Inc. are extensive, starting with the first commercial flux emerald in 1938 and the first flux ruby in 1958. Carroll Chatham was a pioneer in developing the first flux gem in many types, and Tom Chatham and his brother John also made many contributions along the way. These dates are not one-day "Eureka!" discoveries, but a slow and sometimes painful path of evolution that took years to perfect. Opal was accomplished through buying Gilson intellectual property. Everything is measured by saleable results, not laboratory curiosities to write a scientific paper on. Time is money and each day lost in production is dollars lost in sales.

Examples of Chatham lab-grown sapphires. Credit: Orasa Weldon/Courtesy of Chatham Created Gems.

said. "I truly hope you make this big time."

And so it went, Creative Crystals got the jump on Chatham and began advertising heavily in the trade, while Chatham did nothing except the usual gemological talks and donations.

I'm not sure what the downfall was for Dave, maybe money, maybe just because he was a startup. He came from the natural stone trade, so he knew the business, but didn't have our connections in the "created" world. His advertising did pay off for us, however, and we did start to get many requests for alexandrite.

During one of my many trade talks, this one to the employees of Stuller Settings, which sold our products, I was asked repeatedly about alexandrite. Matt Stuller was in the back of the room, and I remember saying, "Matt, I had nothing to do with this..." It was the start of our alexandrite success. When Stuller began to advertise our alexandrite it took off like a rocket in sales and many small retailers followed suit and called us direct for the stone.

Sales were strong and continue to be so. Dave closed his operation and sold off his inventory, but we remained friends. It is only conjecture on my part, but I think it was the pent-up demand for the June birthstone and mathematically 1/12 of the population is born in June, so a built in need was waiting for us to exploit. Proper cutting and orientation are also important factors with alexandrite.

Blue Sapphire (1975)

All corundums are basically the same. When red, it's called ruby. Any other color is called sapphire. After overcoming the problems of ruby, we decided to expand into one of the most popular gemstones in the world: blue sapphire.

This should have been a simple ingredient swap: take out the chromium (for red) and add titanium oxide and a few other elements for blue. The first run resulted in black sapphire, so we cut back on the coloring agents. The next batch was colorless with black points at the end of each crystal. We fooled around with every imaginable concoction we could think of and got nowhere. Five years of research down the drain.

Dad and I decided to try another color and find out if we were just stupid or was there something unique about blue sapphire. (It often does occur color-zoned in natural crystals). We decided to try to grow padparadscha, an orangey-pink sapphire found in Sri Lanka. The true orangey-pink is extremely rare and expensive and often leans more toward just orange or pink sapphire. In the first year of research, we did it! A beautiful orangey-pink color was achieved and became another product from Chatham.

It also told us we were not so dumb with blue sapphire; we just were going down the wrong path. We finally figured out how to grow decent colored blue flux sapphire. We still get some color zoning, but overall, it was a success.

Diamonds (1993)

This was Carroll Chatham's dream, to produce gem diamond in the laboratory. We occasionally played with the process – we had a capable press in the laboratory – but we never figured out the secret to success.

About 10 years after my father's passing in 1983, I began my research in Kyiv, Ukraine, making colorless diamonds with the HPHT (high temperature high pressure) process, which we announced to the trade at the JCK Show in 1996 (see Chapter 31). At the same time, we were also investigating the CVD (chemical vapor deposition) process to make diamond. Apollo Diamond in Boston run by Robert Linares was very close and did have some successful runs but eventually ran out of cash.

Today, we continue joint ventures in CVD diamonds built on the knowledge gained by these early pioneers.

Yellow Sapphire (2000)

Once we mastered ruby, padparadscha and blue sapphire, we began to experiment with all colors of sapphire, even black. This was dad's idea and when he grew it, then showed me the results, I said to him, "Why? There is no market for black sapphire." He did these things just for his enjoyment, but I also told him he was wasting time and money. Each experiment was about $20-25,000 in time

Chatham lab-grown diamonds – blue, green, pink and yellow (0.13 - 1.34 carats) from the 1990s – represented faceted material from Chatham for GIA to study. Credit: Maha Tannous/© GIA.

and materials, so I always felt any experiment required a profitable goal expectation. He would just laugh it off and shoo me out of the labs and back to the corporate office.

After his passing in 1983, my brother John and I continued to explore the many colors of sapphire, including yellow and white or colorless. Sapphire comes in every color of the rainbow and if it was popular in natural, we tried to produce it.

Opal (2004)

Of all the gemstones, opal is the most difficult to grow in the lab. It takes 24 months to produce and is not a crystal, but a type of sedimentation called colloidal chemistry involving silica. I bought Pierre Gilson's opal and emerald processes on the open market 10 years after Pierre sold his company.

I was curious about his emerald process since his stones were identical to Chatham. I believe someone analyzed the trapped flux in some of our stones to uncover the makeup, but the physical setup of his furnace was completely different than ours. Pierre did an excellent job in his emerald crystal growth and the Japanese owners and Kiyoshi Kobayashi further perfected it.

Aqua Blue Spinel and Morganite (2012)
This was just another sideways move in the beryl family that we added to our stable of gemstone offerings. Natural morganite was becoming very popular and expensive. We grew some aquamarine beryl but found it too difficult and switched to spinel in an aqua blue color.

Champagne Sapphire (2016)
Champagne sapphire is very similar in color to morganite but with a slightly different hue; it is tougher as well.

Today, there is nothing new on the horizon in crystal growth for gem use in jewelry. One area I always keep my eye on is the return on investment versus the saleability of a new product.

Take tanzanite, or the mineral zoisite, which my late friend Campbell Bridges discovered in Tanzania. This blue gemstone is only found in Tanzania, so he wisely changed the name to tanzanite. As beautiful as it is, it does not command high enough prices in most stones to create a market for us.

Another is tourmaline, a common natural gemstone, but extremely complicated in its make up. It also has a relative low price per carat in common colors, so it doesn't make sense for us to spend the time and money on developing it.

And finally, jadeite, whose mineral name is pyroxene. It is extremely tough, very hard and very dense. In the right color of green and translucency, it is almost priceless. General Electric created it in HPHT presses and invited me to explore its potential in the gem market.

My conclusion: It must be made much cheaper or be of much better quality.

This 884-carat crystal and three-faceted stones represent the new flux-grown pink sapphires. Chatham's ruby and sapphire crystals became so big — because of the seed size and perfection of the growth process — that it became wasteful to cut away so much gem material. The record is a 2000-plus carat ruby crystal that resides in our corporate office museum shown on page 216. Credit: Robert Weldon.

I also warned them of the following: "This stone will also be sold as natural in China, and be an embarrassment to GE." I advised them not to risk the bad publicity of buyers being taken for large sums of money because it cannot be easily identified. GE agreed and dropped the project.

GE gave me the 10 cabochons samples, which I took to all the experts in Hong Kong and China. No one could identify the stones as lab grown. I kept one and donated the rest to labs around the world, including GIA.

35

Marketing with Integrity

We have donated thousands of our gemstones to research centers all over the world so scientists can better understand their unique characteristics and repeated types of inclusions. In this way, any colored stone can be origin identified because each growth area, be it in the earth or a lab, leaves some birthmark that can be categorized.

The process of deception has been honed throughout the centuries of the gem trade.

I have had some experiences in the ruby deposits in Chanthaburi, Thailand, miles out in the bush. I was not cheated but met people who were, and this is how it happens.

It is easy to spot a dig: a tripod of wood poles with a pulley attached over a 4-foot round hole. It's hot – hotter than anything you can imagine – and you're all dressed up in your finest Banana Republic bush clothing, sweating like a stuck pig. The miners are wearing only a loin cloth, and are very tan, gaunt (half of their teeth are missing) and miserable looking.

A basket comes up from 20 feet below with dirty gravel in it. The pulley man grabs the basket and dumps it on a screen and begins to sort through it. This is repeated several times until one basket comes up with something red in it among the gravels. It is poured out on the screen and the red rounded pebble is picked out and the miner smiles a big toothless grin. The stone is about 10 carats in size, all roughed up and unreadable, the shape is correct for rough ruby that has been tumbled in river gravel for a few thousand years and the color is superb.

"How much?" you ask. They shake their head as if not understanding. OK, you think to yourself, we have a language problem here, not uncommon, so you use the only universal language that works worldwide – you start peeling off bills of the local currency. Depending on the situation or how gullible you look, a show is made of counting bills. Maybe $500, maybe $5,000.

The miner scoops up the bills looking at each as if you're a potential counterfeiter (this cements in your mind his caution and inexperience), nods his head and hands over the 10 carat pebble that you think will cut out a 6-7 carat stone that could fetch 7-10 thousand US Dollars per carat!

You get back into Bangkok and look up a cutter – there are thousands of them in the town and surrounding areas. The cutter does not ask questions about the source of the stone or the price; it is none of his business. A finished price per cut carat is arrived at; the price ensures the cutter will not waste rough and that you will get a poor lumpy cut at $100 to $200 for cutting. It can always be cleaned up in New York.

The cutter delivers the cut stone, and it is a beauty: eye-clean and bright-red, every stone buyer's dream come true. You take it to the Asian Institute of Gemological Sciences (AIGS) a well-known and respected Bangkok gem lab, that tells the truth about origin. Not value, just identification if natural or not. They have seen just about every trick in the book and that book is thick! You need certification to get top dollar for a stone, especially in Bangkok.

The report comes back, you're now down another few hundred dollars in cost, but no problem, you think. The report, however, says "synthetic corundum, flame-fusion." (The cost of this material is 10 cents per carat!) Or it could come back as "flux-grown, synthetic ruby maybe Kashan or Chatham" and maybe worth $500, but that was a rarity. The miners in the fields don't spend huge amounts to cheat people, just pennies.

This is a case of buying something based on your "gut" feelings and assumption. After all, you saw the stone come up out of the mine, didn't you? Well, yes, right out of the guy's pocket down in the hole!

Again, another perverted sense of "honesty" in the gem business. And if you go back to the mine and complain you would be accused of switching stones and trying to cheat them, and be told, "Get away or we will kill you!" This is a very real possibility out in the middle of Thailand.

There are no returns in the stone trade, especially in the mining areas. This is the way it often goes down: you're in the area where precious stones are traded and someone comes up to you with a bundle in their arms. "Do you want to buy stones?" they ask as they unfold the wrapping. There sits a great big green rough stone. "How much do you want?" you ask. No answer. "Is this a natural stone?" you ask. No answer. This stone is big, like a golf ball.

I have seen this in Bangkok, in Africa and another in Colombia. You pick up the stone, look for any tell-tale inclusions or giveaways. One I saw in Africa, hundreds of miles in the bush, was the scariest. I said, "I will give you $100 dollars for the stone." My buddy, Abe Suleman, an accomplished stone dealer from Tanzania, stutters, "Tom, are you trying to get us killed?" (This can happen and has, when one has used bad manners or calls someone a crook.)

"No Abe, this is the best fake emerald I have ever seen, and I want it for my collection," I said. "This is glass, doctored to look like a natural emerald in the rough."

The dealer said nothing but understood me. He has taken a chance with his life by trying to sell a doctored piece of glass. He rewraps his stone, puts it away and walks off, without a word.

If I offered $10,000 or $100,000 for it, a reasonable sum for such a piece of natural rough, he would have taken it. And once we tried to cut it, and in one case we did, it cuts like butter, it is so soft, clearly indicating it was not natural. If a buyer offers $100,000 for a green rock, that is his business. The seller has done nothing wrong, in their perverted sense of honesty, because he never said it was a natural emerald, you assumed it was. Of course, it was misrepresented by its doctored appearance that made to look like a natural emerald, but this ain't church folks, this is a look-you-in-the-eye reality and if you screw me, I'm gonna kill you, sort of game!

The Chatham Created Emerald made this even more possible, and it has happened, for huge sums of money.

Back in the late 1970s, a middle-aged gentleman asked to see me in my San Francisco office. Our doors were always open at corporate (versus the labs), so he was welcomed in. I could tell he was agitated; I kept a 9mm pistol in close range under my desk, just in case.

"What's on your mind?" I asked. He pulls out a sheet of paper and it identifies itself as a lab report from the government of Bogotá, Colombia. It pictures two emeralds, 9 x 7 emerald cuts, with all the characteristics listed down to the exact weights of each.

"Looks like a gem report, what's on your mind?" I ask. He then pulls out a stone paper and unfolds it to reveal two perfectly matched, 9 x 7 mm emerald cuts, of gem quality and color. "Nice stones," I comment. "What's the purpose of your visit?"

"I paid $50,000 cash for these two stones," he said. "GIA says you made them; they are synthetic! "FAKE" And it's your fault I was taken!"

I picked up the paper of stones, two perfectly cut stones of about 2.5 carats each, a perfect match, a miracle in the natural stone trade. "And your point is what?" I asked.

He ranted and raved about the injustice and criminal activity of my business and how I was responsible for his loss. In his mind, we should not be allowed to exist!

I picked up his Bogotá government certification and said, "First, this is a piece of worthless garbage, not worth the paper it's written on. Bogotá does not have a gemological laboratory like GIA. Second, stones of this quality and matching size would be worth far more than $50K, easily $100K. Do you think because you're a white guy and wear nice clothes, that you're smarter than some poorly dressed peasant who is hanging around the trading area of Bogotá?

"What is your gemological training?" I asked.

"None, I just know my gemstones," he sheepishly stated.

"Not too well, from what I can see," I said. "What do you do for a living? Not buying stones, I hope!"

"I am a pilot; I fly commercial airplanes all over the world," he said.

"Well, if you had the proper training in gemology, this would not have happened," I said. "I wouldn't dream of flying a plane without extensive training and you should stay out of a business you have not been trained in! Goodbye, and don't blame me for your uneducated mistakes."

He was not a happy camper.

Chatham has not and will not support this type of chicanery. We have donated thousands of our gemstones to research centers all over the world so scientists can better understand their unique characteristics and repeated types of inclusions. In this way, any colored stone can be origin identified because each growth area, be it in the earth or a lab, leaves some birthmark that can be categorized. That knowledge is spread throughout the world as an indicator of what to look for in identification.

But you still must read the books, take the courses and keep up with the scientific developments. It is not rocket science, but it does not work by osmosis. Some people never learn.

{199}

36

How to Travel Safe and Work the Trade Shows

Our messages were clear and to the point: Chatham Created Gems are real gemstones, identical to their natural counterparts and available in well-cut calibrated sizes. We also emphasized that they were much cheaper than their natural counterpart, much to the chagrin of some natural dealers.

There were only three ways to reach your customers when I started in the wholesale jewelry and gem business: on the road as a salesman, through trade advertising and in-person at trade shows.

A traveling salesman spends his or her time on the road going from store to store to show jewelry pieces from a designer or manufacturing company. Unfortunately, organized gangs, often trained in Colombia, saw these salesmen as easy targets and a relatively safe way to make a living. They were very creative and often worked in teams of three or four people. In the beginning, it was simply a matter of following a salesman's car to a gas station or hotel after they left a store with their wares. Then, thinking they were totally unknown to anyone, the salesman would stop for lunch or a coffee, leaving the case of jewelry in the locked car. It didn't matter if the goods were on the back seat or in the trunk, in about 10 seconds the thieves would smash the window or pop the trunk with a crowbar, grab the jewelry and be gone.

Insurance companies were not covering most of these hits because they considered it "abandonment" on the part of the salesman. Many jewelry companies went out of business as a result of these robberies. New rules were implemented as the trade became more aware of security, requiring salesman to be with their goods 24/7. Nothing short of a pistol to your head was deemed a justifiable insurance claim.

The thieves, in turn, became more creative and brazen. In airports they would squirt mustard or ketchup on your back, then offer to help when they made you aware that you had something on your clothes. Naturally, the salesman puts down his jewelry case and takes off his or her coat. Two or three people would be involved and distract the salesman while another made off with the salesman's bag. It was so slick the thieves rarely got caught and the original "helpers" were still standing there looking innocent.

It has happened in airports right in front of me – and I didn't even see it. One example was in Las Vegas at the JCK show. An Asian woman had three matching bags in front of her, but because jewelry is heavy, it was easy to guess which bag the jewelry was in. She was maybe three people in front of me when all the sudden she starts screaming, "My bag…who took my bag?" and frantically looks all around us, even behind the ticket counter. It was gone, right before my eyes.

{201}

The Chatham organization was hit three times and caught them in the act each time – but there is nothing you can do because they just say, "Oh sorry, I thought it was my bag." The last time it happened was at the LAX airport where we were flying in for a jewelry show. The thieves knew when and where all the shows were taking place because they were advertised in the trade books. It was easy for them to spot people with carry-ons with possible jewelry in them. They didn't always select correctly and sometimes just got dirty laundry, but they were very successful nonetheless.

I was waiting at the luggage pick up area with two carry-on suitcases next to me, both loaded with jewelry and gems and the handles fully extended, one on each side of me. The luggage started to arrive, and everyone shuffled around, as they do. A well-dressed Asian woman casually walked by me, tipped the handle of my carry-on and proceeded to wheel it away. I was keenly aware of the exposure and very alert, however, so before she got 12 inches, I firmly grabbed her wrist and pulled her back. "Let go of the bag, lady," I said staring her down. She immediately let go of my bag – she knew that I knew what she was doing. "Oh, so sorry, I thought that was my bag," she said as she walked away. There was nothing I could do, nor did I think she could be charged with anything.

Our company started using Brinks to transport anything of value.

The next way to reach customers was trade advertising, and I had a ball with that. Rates were very economical in the 1970s and 1980s: a full-page, four-color advertisement ran about $1,000 per full page for a 12-time rate plus production costs. Our messages were clear and to the point: Chatham Created Gems are real gemstones, identical to their natural counterparts and available in well-cut calibrated sizes. We also emphasized that they were much cheaper than their natural counterpart, much to the chagrin of some natural dealers.

I was always looking for ideas from other industries that I could apply to our products. One day, my wife and I were in a supermarket shopping, and I noticed that shelf after shelf had either a can or a package saying "100% natural…," but then I would read the labels and see all the stuff that went into the product. I realized that there really isn't anything "un-natural" on this planet, so these companies were playing a word game. Maybe everything in the package was "natural" but I doubted they formed naturally as this one product.

Since the gem trade was obsessed with "natural gems" when they really were quite doctored and treated. I thought I would have some fun advertising our goods as being 100% natural. My concept ad was a sharp jab at the natural trade's side: it featured a bowl of emeralds under the caption: "100% natural ingredients" and on the side of the box I listed what was in the product.

"Each Chatham Created Emerald contains the following:"

100% natural Beryllium

100% natural Alumina

100% natural Silica

100% natural Chromium

The full-page, full-color ad ran in *National Jeweler* magazine. Half the trade had a fit and objected; the other half loved it! Maurice Shire, a natural emerald dealer from New York, went ballistic. I also got a call from Sue Patch from the FTC. We knew each other, but we were on opposite sides of the issue of created versus natural gemstones.

This was, in my opinion, the most effective advertisement Chatham ever created. The industry was evenly split on its opinion, but everybody got the message: A Chatham lab-grown emerald was not an imitation emerald! The FTC asked me not to run it again but admitted it was not in violation of any FTC guidelines. It was reprinted in most industry magazine as a news story. Money well spent!

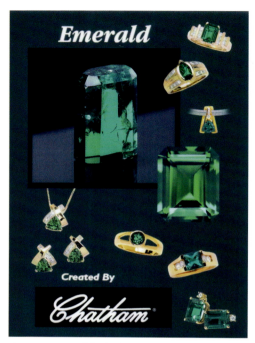

A very successful Chatham lab-grown emerald advertisement, before and during the shows.

"Hi Sue, what can I do for you?" I asked, as if I didn't know.

"Tom, my phone is ringing off the hook with calls from irate stone dealers over your cereal box ad," she said.

"Sue, are you telling me the ad is illegal or misleading?"

"No Tom, I am not, but I am asking a favor. Don't run it again. Please."

"Well Sue, I only planned on running it once, but if you want, you can tell the callers you got me to stop the ad and get them off your back."

"Thank you, Tom, much appreciated."

The ad caused such a stir in the industry, however, that all the other magazines in the trade re-ran it as a news item with all sorts of comments, most positive. It was the best ad I have ever created. To keep the ball rolling I had thousands of small cereal boxes made and handed them out at every trade show for the next two years – with green Jelly Belly's inside. Almost everyone loved the idea, and it created a lot of awareness for our created gemstones.

Traveling salespeople were not our answer and trade ads, except for the cereal box, are ineffective, in my opinion. It took 10 years to get the message out that we grew ruby and other gemstones and sold them as cut gemstones. People would walk up to our booth at a trade show and remark, "Oh, I didn't know you sold loose stones, I thought you just made those crystal clusters." That drove me crazy: 30 years of advertising down the drain!

So, the third way of reaching the trade – and the best way, in my opinion – was trade shows. I embraced them 100 percent. Unfortunately, so did everyone else. All the major shows were full with waiting lists a mile long. Adding to this was a favoritism that took place in every venue: trade show "managers" were wined and dined, and some received free jewelry to get a spot in the show. It was really a "who-you-knew" business. Most of the shows were held in hotels, ballrooms and even individual hotel rooms.

As the popularity of trade shows grew in the United States, however, so did the venues. Hotels no longer had enough room, so convention hall venues started to be used. This also eliminated the "special" positions held by show managers because they were dealing with city bureaucracy instead of hotel managers who could be open to a kick back.

When this change took place, we were still trying to get into the major trade shows with minimal success, but by the mid-1980s we were in at least 25 shows a year. Attending so

{204}

many shows was a grueling challenge, with trying to run a business, keep a marriage together, raise kids and constantly be on the road.

I probably missed a few that were bad because nobody attended. There is nothing worse than a dead show, but you never know until you go. I would often pre-shop these shows as an attendee just to get a feel for it. You get a vibe from a successful show.

One year, I tallied the days I was out of the office for an IRS audit on my auto expenses. Total: 250+ days gone. The agent said forget it. That left less than 100 working days to squabble over and didn't count the attendance of gemological conferences or speeches I gave all over the world!

We never sold anything when we first started doing shows, we just handed out brochures that included price lists and ordering information. This was not too unusual because all jewelry shows were show and tell only, with potential buyers stopping by booths to see what was available. If the retailer couldn't see it, they would not order it. Delivery of jewelry was not allowed at the shows for security reasons and because the items were made based on orders from the shows. Of course, some deliveries did take place, but it was the exception, not the rule.

Jerry Hulse and I did most of these shows, which were relatively inexpensive, about $1,000 per booth, plus hotel and airfares. They later became very expensive and required more booth space with special displays. We evaluated the first shows based on how many came up to the booth and took our brochures. Later on, we counted sales by the day!

It was the complete opposite experience at the Tucson Gem shows, which were like no other in the world. These gem shows covered at least

Trade shows we participated in:

The Pacific Jewelry show,
San Francisco – once a year

The Pacific Jewelry show,
Los Angeles – twice a year

The Oregon Duck show,
Portland, Oregon - once a year

The TOLA show,
(Texas, Oklahoma, Louisiana & Atlanta),
Dallas – three times a year

The Jewelers of America show,
New York City - twice a year

The Jewelers of America show,
Chicago - once a year

The Jewelers of America show,
New Orleans - once a year

The Cincinnati, Ohio show - once a year

The Tucson Arizona shows in
February, with three locations - once a year

The Jewelers International show,
Miami - once a year

The Jewelers Circular Keystone show,
Las Vegas - once a year

The Independent Jewelers Organization
show, which moved to a different state
every show - twice a year

The Retail Jewelers Organization show, one
in Des Moines, Iowa, and one that moved
every year to a new state - twice a year

The Canadian Jewelers Association show,
Toronto - once a year

The Hong Kong Jewelry show
in September - once a year

The Bangkok Jewelry show,
Thailand - once a year in September

{205}

We strived to grow large single crystals for cutting gemstones, and although these crystals look beautiful, most are not of good enough quality to facet. Many of these crystal groups sold out yearly at the Tucson shows.

30 venues – from a big convention center to gas stations and hotels renting out space – and some of the smaller shows ran for 30 days. It is a dusty and dirty and exhausting experience. All the hotels triple rates for rooms and food and the rental car companies usually ran out of cars.

Tucson was well worth the effort for us, however, because we could sell anything the labs made that did not meet our criteria for the established jewelry trade. Rough emerald so poor in quality it was uncuttable, but we were able to sell it.

These shows were not organized to start at the same time and some intentionally jumped the gun and tried to open a few days before the main, bigger shows. People only had so much money to spend and so many days to wander around the huge tents erected outside of old motor hotels (the beds were shoved in the bathrooms so a showcase room could be set up) This was a real "buyers beware" event and most was cash and carry. If you liked something, you should buy it on the spot, or it will be gone – or you wouldn't be able to find it again.

I heard about the Tucson shows through the trade books, but they overlapped with our big show in New York. One year I took off a day early and stopped in Tucson to check it out. I always visited a show before I would join it, especially when booth prices started to climb (up to about $5,000 for one booth in 2020). We had grown considerably so we needed at least two booths and at some, such as at the JCK Plumb Club, we rented eight booth spaces.

The year I stopped off in Tucson was about 1975. I went to the main show at the Holiday Inn on Broadway and was amazed. The shows were supposed to be strictly wholesale, but I could sense otherwise. There is no law against selling retail, but it really ticks off your retail stores if you do, and I can't blame them.

Your Guide To Profitable Buying

at the RJA International Jewelry Fair

RJA's Annual Convention and International Jewelry Trade Fair comes at the end of this month. Plan now for full use of all four opportunity-packed days at the show. JC-K's annual pre-convention program and buying guide makes planning easier. It gives everything you need to know *before* you arrive at the New York Hilton and Americana Hotels on July 29. First, tear out the fold-over *Buying Guide*. Study the condensed program on the back page. Then, read the full program rundown inside. These descriptions will help you pick the seminars and convention sessions which most interest you. Next, examine the alphabetical list of exhibitors. Check off the ones you wish to see. Now, turn to the charts which show exhibit areas at both hotels. To make your planning easier, these floor plans show both exhibitors' names and booth numbers. Mark the booths you wish to visit. Finally, look at the pictures of new products to be introduced at the show. A booth number is given below each picture. When you see an item you think will sell well in your store, mark the booth number on your floor plans. Now you are ready to plan your days at the show. Use the floor maps to plot a fast, efficient walk through each exhibit area. You'll have time to visit every booth on your list and attend many seminars.

Trade shows were very important to the industry – and even more so to Chatham because, unlike many manufacturers, we did not employ traveling salesmen to show our stones to retail stores. As the trade shows became bigger and bigger, they eventually outgrew even the biggest hotels and moved into convention centers. "Your Guide To Profitable Buying," was published by *JCK* magazine in July 1973 to help buyers navigate the ins and outs of jewelry shows by mapping out a plan. Today, major shows are 10 times the size of the earlier shows and require a team of buyers all moving with missions in mind – not idle wanderings down aisles.

There is a wide gap between consumers and the trade, and there are rockhounds, hobbyists, craft people and the friends they bring in with them in between. Because we were selling mostly rough gemstones, (uncut) we had little appeal to a consumer, but the cultured pearl booth could sell anyone. Many marked up prices because of this infiltration and only when you were vetted did you get the wholesale price. You had to show a resale license to get in, but it might have been a license to sell furniture! It was (and is) very loose in Tucson.

Because of this, we always wore our wholesale exhibitors' badge when visiting other venues in Tucson, which we always did. The badge spoke for itself. You couldn't get anymore wholesale than we were!

Once again, the "Legend of Chatham" paid off and I met the organizer, Warren Matthews, a typical rockhound who loved his whiskey and rye. I could smell money in this show because it would make it possible for us to sell our seconds and rejects from the lab – and we had barrels of that in storage!

I asked Warren if I could be in his show at the Holiday Inn, knowing it was sold out. I knew if Warren wanted you in his show, he made room. The show ran from one weekend to the next – nine days long – and from 9 a.m. to 9 p.m. Luckily for me, the hours were changed after my first year to 10 a.m. to 6 p.m., which I thought was much more reasonable. It was a laid back and relaxed atmosphere, not an uptight, suit and tie sort of affair, dealing with top executives from big store and chains. The hours were grueling, but we still had fun and met people from all over the world. In total, we did the Tucson show for more than 45 years.

At first, I got a 6-foot-long by 4-foot-wide booth in the main ballroom for about $4,500. I complained to my neighbor that her beads

were falling into my booth, and she quickly informed me that "the reason I am short of room is because Warren carved your booth space out of mine, with no reduction in price, so don't complain!" That's what legend buys, but I was not very proud of that situation.

We soon expanded to two booths and then a new show opened in a big tent across the street, so we took a few booths there too. And sales doubled! It was strange that sales would expand with the same goods in a booth only 300 yards away, but it worked. You missed people, people got lost, people went to other shows. I never knew why, but it worked.

During one of the early Tucson shows, Warren, the show manager and owner, went into one of the show rooms and got into an argument with a dealer who told him to get away from his booth. "I can go any damn place I want in this show, I own this show!" Warren said. By the end of this day, many dealers had organized a sort of a union and proposed a new show, run by the dealers.

As was typical in many shows, halfway through you had to commit to the next year's space with a 50 percent deposit. One of the dealers came into the sign-up room and shouted, "Don't sign a new contract, we are starting a new show."

Warren had him arrested and thrown out, but that was when the American Gem Trade Association (AGTA) was formed. The dealers started a competing show just down the street in the convention center, which was a much larger venue. I didn't like their approach of coming into Warren's space to recruit members, so I didn't join the AGTA. If I had, it would have saved me $250,000 in future court costs!

Warren had no trouble filling the spaces left

{208}

by these dealers. Tucson was becoming more and more popular and was growing right and left. You could find geodes big enough to stand in or a complete T. Rex skeleton, plus gemstones from every corner of the world. It was truly a cornucopia of everything found in the earth. Slabs of quartz for tables, petrified wood slabs, you name it, you could find it in Tucson. "Going to Tucson" became a must for the rockhounds and those in the jewelry trade.

There came a time when Warren was losing his lease on the old Holiday Inn. It was falling apart and stands empty to this day, condemned. He tried different venues, but none were in the center of the action, like where AGTA had landed. So, I went to AGTA and asked for booth space. "Nothing available" I was told, but then I could see new companies coming in that didn't even exist when I first asked, so I went back to the president of AGTA.

"I want a booth in your show, and you have no grounds to deny me," I said.

"We don't want you in our show, Tom," the AGTA president at the time said. "No created stones allowed. Everything is natural stones, so you don't qualify."

"But you have booths selling cultured pearls; you have existing members selling created stones, so I want equal treatment!" I responded. "I don't care if you don't like my product or me, you are selling to my customers, and I want equal access."

The answer was "NO, we don't want you."

Finally, after years of being put off, I told Owen Borderlon, then the president of the AGTA, "Owen, I want in and if you refuse, I will take legal action."

"Take your best shot, Tom. We won't let you in," he said.

So, I sued the board of AGTA for antitrust and conspiracy to restrain my trade. I sued all the board members and served them publicly at the JCK show in Las Vegas. No one had the proper insurance for this sort of lawsuit, nor did the AGTA. Since I was a San Francisco company, all court appearances had to take place there and their attorneys had to fly in from New York, Texas, and a few other states. To say they were furious was an understatement and to add even more trouble, one of the defendants had a New York lawyer who kept making motions that all had to be addressed in court, and all the attorneys had to fly in each time. Our attorneys were based in San Francisco, of course, but some months the tab ran up to $50,000!

After a few months of this, Owen asked for a meeting. Things were not going their way and they knew the outcome was obvious; they had conspired to keep me out of their show.

"Tom, we are going broke. We cannot continue this fight and we are willing to let you in the show if you drop the lawsuit," he said. "Owen, this was never done to put you out of business or make money in a settlement. All I wanted was equal access," I told him.

Case closed. Or so we thought. Then all the board members sued the AGTA for legal costs. I am not sure what happened with that action, but I think there was some repayment made.

The next year we joined their show, placed on a back wall. Due to all the press the legal suit brought, we had a great show and continued there for many years, in that same location, with the same size booth. I knew that was not going to change, ever. Many in the AGTA were very upset with the actions I took. Some were good friends, before, but they had no idea what we went through to get where we were. It took some of them years to get over the fight. AGTA ended up buying the proper

LETTERS

Chatham Takes Issue With AGTA Board

SAN FRANCISCO—In the October 16th issue, p. 31, "AGTA board recommends adoption . . . " a biased and incomplete picture was portrayed by the writer, Mary Ford.

I am disappointed in the reporter's lack of journalistic objectivity by not explaining all facets of the issue and using our company name on top of it.

The AGTA does not represent our firm nor the colored stone industry. A paid membership of fewer than two hundred is hardly a majority of colored stone dealers.

The AGTA gives the impression of unification and authority in the colored stone industry, yet prohibits competing companies like Chatham Created Gems from joining this "single voice" group.

The AGTA claims it is promoting colored stones for the benefit of the industry, yet it advertises and sells directly to the public and private investors. Even the JA show gives the AGTA special treatment, away from the real trade, to service their members' "private accounts."

The AGTA claims it is for truth in advertising, yet it cannot reach a decision on disclosing artificial color enhancement.

By supporting and suggesting the adoption of a foreign country's rules, the AGTA is, in effect, saying that it does not support U.S. federal law—nor the years of work accomplished by the Jewelers Vigilance Committee—nor the American Gem Society, in regard to the precedented (by U.S. courts) nomenclature rules and guidelines governing man-made stones.

The article suggests that Chatham Created Gems is in violation of such rules, and that Chatham Created Emeralds should legally be called artificial emeralds. Under the decision handed down in the Federal Trade Commission's hearings of "Chatham vs. the U.S. Government," 1963, Chatham is in full compliance with the federal law. The article implies we are not in compliance. Furthermore, it is an attempt to inaccurately describe our product with the intent to confuse, defame, and in effect, restrain our commerce.

There is a faction in the colored stone industry that has actively undermined the efforts of created gemstone producers. Misinformation in trade journals, proven concepts of technology and chemistry ignored and the misuse of terms such as fake, imitation, simulated, artificial, synthetic, etc., are purposely misused to confuse and discredit our efforts to properly educate the trade.

Chatham has weathered battles with the Federal Trade Commission of the U.S., and was given a clean bill of health. We have gone head-to-head with the Jewelers Vigilance Committee and have

felt the American Gem Trade Association (AGTA) was discriminating against Chatham by refus-
ng to allow us to exhibit at its Tucson Gem Show held every February. It claimed exhibitors could
only sell natural gems, but many cultured pearl companies, a very manufactured product, were
llowed. I wrote this editorial objecting to what I felt was an inaccurate portrayal of Chatham in a

come to mutual agreement—and a seat on the board. We have survived petitions to ban our participation at American Gem Society conclaves, only to be admitted as registered suppliers, and we have been disallowed membership in an organization (AGTA) purportedly representing the colored stone industry. Now we must fight off foreigners camouflaged with the "universal language." Funny—last month I was in Hong Kong, Bangkok, and other gem centers in the world, and English seems to be the common language—not French.

The anti "created" stone faction has now retrenched via the AGTA and is again pushing the same ideal that was previously rejected by the FTC and JVC and not followed by the American Stone Importers Association.

This rejected rhetoric has once again come forth under the auspices of a national publication because ". . . . AGTA board recommends adoption . . . "

The AGTA, by implication, is ignoring both U.S. federal laws and the very foundation of the Jewelers Vigilance Committee.

I view this as a planned attempt to restrain our business and an attack on the character of Chatham Created Gems, Inc.

Created gem producers are due a correction in the form of a full explanation. The AGTA should be explained for what it is and is not. CIBJO, its origin and explanation should be given. And last but not least, a look at why the FTC and the JVC do not endorse or support the CIBJO rules. All of this is due the created gemstone producers in general, and Chatham Created Gems, in particular, and it should all be done with equal conspicuity.

—*Thomas H. Chatham, President*
Chatham Created Gems, Inc.

Editor's note: NATIONAL JEWELER *did not state nor mean to imply that Chatham Created Gems is not in accordance with any U.S. guidelines, regulations or accepted industry practices.*

Joel Windman, executive vice president, Jewelers Vigilance Committee, said of the rules in question, "CIBJO rules have no bearing on U.S. commerce." He noted that in cases where the rules conflict with federal, state or city laws, U.S. laws take precedence.

Moshe Pereg, president, Jewelry Manufacturers Guild, said that AGTA does not have the authority to set guidelines for the colored-stone industry. He said that there are enough laws regulating the industry and protecting consumers, although lack of enforcement is a problem. He suggested that AGTA should concern itself with monitoring its members and working to eliminate problems in the appraisal and investment areas, which he believes could severely hurt the industry.

National Jeweler welcomes comments and opinions from readers. Please send letters to: Editor, National Jeweler, 1515 Broadway, New York, N.Y. 10036.

---

was an active contributor to many jewelry publications, writing in defense of laboratory-grown insurance to protect itself from any action like mine from ever happening again at their expense. Eventually, AGTA made me a full member, not just an associate member.

At one time we had three booth locations in Tucson: two booths in the old Holiday Inn, one in the AGTA convention hall and three in the Gem and Jewelry Exchange (GJX) tent. It took a crew of at least six people to run these booths and we sometimes had to have three people in the GJX tent because of our space size, a total frontage of 30 feet of exposure with a lot of crystals lying out on the table.

There are a lot of distractions in Tucson, with a lot of old friends from around the world stopping by to say hi. People I would never see again except for this show: Daniel Sauer from Rio; Julius Petsch from Idar-Oberstein; Alan Hodgkinson from Scotland; Campbell Bridges from Africa; Fred Ward from *National Geographic*; Richard T. Liddicoat and Dr. Jim Shigley from GIA; Kennedy Ho from the Asian Institute of Gemological Sciences in Bangkok; Paton Kelly from QVC; Alan Bell, CEO of Rio Grande in New Mexico; Matt Stuller of Stuller Settings; Lorenzo from Enzo in China (with 1,000 stores); Dr. Fred Poe, a longtime supporter of my father's work; Helmet Swarovski and Dr. Eduard Gubelin from Switzerland; Robert James from the International School of Gemology; Antoinette Matlins. The list goes on and on and I know I have missed a lot of old friends.

After 45 years, the entire gemstone world has come to Tucson at one time or another and they all stopped by my booth to say hi. It was very gratifying after so many years of friction. Even Maurice Shire stopped by one day and I asked him a question about some unusual ruby crystals we had on display.

"Maurice, I know you feel that everything grown in a lab should be called 'synthetic,' but what about these platinum crystals growing on our rubies are they synthetic? Because if they are, I should be awarded the Nobel Prize in chemistry for creating one of the basic elements, a feat unaccomplished to this day!" He just walked away.

We had a few thefts. Probably the worst was during our last show where I had some beautiful emerald crystals, maybe five or six on a side table, about 200-300 carats each and worth $10 a carat, so worth more than $10,000. I remember that a couple sat in front of the table looking at other stuff. We normally didn't assist buyers as they select their stones, but just waited to be called over to add up purchases.

In my walk around the booth, I looked at the tray of five emerald crystals and noticed that the tall and skinny one with the inclusion on the top, was missing. No doubt during one of their visits, the couple palmed it and there was no way to catch them because it was discovered missing way too late. Thousands of people attend these shows and even if I saw them, I would not have proof that they stole it. After that incident, I put those crystals out of reach in a show case with jewelry. Live and learn. Many exhibitors have had similar experiences, unfortunately.

Believe it or not, I have a connection with a lot of stuff we sold at Tucson. I remember the disaster that created the flux flub rubies, the weird gray crystals with a color change as a result of my father putting beryllium oxide in a ruby run "just to see what would happen." I scolded him for that, "Dad, you just blew $30K in product," after the run was over and these strange crystals came up from the labs to the corporate office. But we sold them all in Tucson! My father did not repeat the experiment.

I learned to bring all these almost worthless crystals to Tucson from a lesson I learned years

{212}

ago when I was just starting out in Tucson. A booth across from me was in the Tahitian cultured pearl business. They brought a large barrel of oyster shells, still slimy and stinking, to every show and offered them for $20 each. He never failed to sell out. They were worthless curiosities, but proved the adage, there is a customer for everything – or words to that effect.

One thing I had fun doing at Tucson was to buy something from a booth down by the freeway (sort of a low rent gem area) and resell it in my booth for twice the price. Obviously, I was getting bored with Tucson. It was a love-hate relationship: up at 6 a.m., dress, eat and be setting up by 9 a.m. It was hard work, and the costs began to get out of control, too. Each bare booth was $5,000 plus the rental of tables and showcases. That was about $30,000 for three locations. Then there are 6-8 people flying in and staying at a decent hotel, another $20,000. We never skimped on food, so dinners ran another $10,000. A grand total of $60,000!

"Tom, this is crazy," Harry Stubbert, who hated Tucson, finally said to me. "Yes, we make a profit, but at what expense?"

We also didn't get new customers because we never showed our top-end line of jewelry in these shows. The biggest reason for quitting Tucson after 45 years, however, was a lack of the flubs and crystals we had stored and built up over a 60-year period. Our production was getting very good, and we produced less and less of these anomalies, which sold for 10 cents on the dollar if we could get it.

Plus, I was getting tired!

So, in 2018 we put up big signs about leaving the show, but not going out of business. We had a record show, and at the end I gave away all the equipment I had been storing

[Tucson] was a love-hate relationship: up at 6 a.m., dress, eat and be setting up by 9 a.m. It was hard work, and the costs began to get out of control, too. Each bare booth was $5,000 plus the rental of tables and showcases. That was about $30,000 for three locations. Then there are 6-8 people flying in and staying at a decent hotel, another $20,000. We never skimped on food, so dinners ran another $10,000. A grand total of $60,000!

every year in Tucson – maybe 30 lights, 20 cords, table lights for diamonds, trays, a back drop, carpet – you name it we gave it away, and shipped one trunk of goods back to the office instead of storing eight trunks of show equipment.

We really hated to say goodbye to Tucson, but never regretted it. We revisited Tucson as buyers in 2020 and reaffirmed our decision to quit. Walking the shows was even more work. I would bet there are 50 different venues, so bring your tennis shoes and leave the heels at home!

37

The Future of Chatham

Once, a group of New York diamond cutters asked to cut a few of my diamond crystals … "It cuts just like natural diamond!" they said. "Well, what do you expect, it is diamond!" I responded. That is the hardest thing to get through people's heads: If I say this is an emerald, then it is. If I say this is identical to natural diamond, it is a diamond! People just have a hard time with chemistry, I guess. It's not magic; it's not alchemy. It's just plain inorganic chemistry.

Harry Stubbert had taken on more and more responsibilities at Chatham Created Gems, Inc. by 2010. Not really taking them – more like creating them. He wrote a five-year business plan that charted his vision of the future. He became more involved with the office personnel and what was expected of them and developed marketing plans and sales pitches for our salespeople to offer to our customers. All of this was new to me and our employees; I was always a seat-of-the-pants type of manager. Harry brought us structure, purpose and direction.

I was also cutting back on my travel and responsibilities. Trips to Shenzhen, China to design new jewelry was not my forte, so I was replaced by Monica McDaniel, a GIA graduate with a flair for design, she is now a vice president in Chatham. Monica is also in charge of all internet marketing, an area growing by leaps and bounds all over the world. Monica grew up with computers; I did not. She also redesigned our website, chatham.com, so it is much more functional and better at explaining what Chatham Created Gems, Inc. does. We are the world pioneers in laboratory-grown gemstones – and she makes sure everyone knows it.

Those 16 to 20-hour trips to Asia really take a toll on your body and if I wasn't needed, there was no sense in going. I don't need to visit the cutting factory, which is managed from Hong Kong. The main factory in Shenzhen has more than 200 employee and is well-structured and under control. If I go it is more like a social call to old friends. (Since the pandemic, both Hong Kong and China are off limits, unless you like 14-day quarantines for both countries. Everything must be handled by computer and ZOOM.)

I occasionally go to Hong Kong for the Hong Kong Jewelry show, especially after we started a new emerald production lab using the old Biron process. We enlisted the help of Artur Birkner, the inventor of the process who almost sank Chatham back in the 1990s during the emerald wars. I came to find out it was not his doing. He had sold his business and the new owners went down a crooked road to oblivion. Artur lost as well, with a lack of pay off, so we enlisted his help and guidance in the new factory to grow emerald by the Biron process, the Gilson processes and the Chatham processes. This was not an easy endeavor, and I was eager to help and give advice when called on. We had several mishaps with explosions, but no one was hurt. Today, all is running smoothly in all growth divisions.

Harry Stubbert

Harry is fully in charge of all daily operations and owns the company 100 percent. He has been a Godsend and a true friend to me and my wife Dianna. Now, at 76 years of age, it is time to pass the responsibilities and headaches to someone I know will accomplish a lot in the coming years. All operations are run from the San Marcos office because our location in the Tiffany building on Union Square in San Francisco became too expensive. Since Harry lives near San Marcos, he can hire new employees from nearby GIA in Carlsbad as they graduate, so it was a good move. All marketing, both in trade shows and the internet, are also run from that facility.

I still hold the title of Chairman of the Board, so I am not out of the loop, but I stay out of the company's operations unless asked for comment. During an IRS audit that was way over the agents' head when it came to understanding what was involved in growing, cutting and making jewelry, for example, the agent asked me if I was afraid of someone stealing a stone or materials because we were using gold and platinum. I answered no because I have no real control over them.

If I do not pay for the work – cutting stones or making jewelry – production stops and we go broke. On the other hand, if they goof off and produce bad cutting or poorly constructed jewelry, I will not pay and will replace them. Do the cutters all have a stone or two for their wife or girlfriend? Maybe, but highly doubtful. If you get caught stealing, you are fired and black-balled in the entire industry. Good paying jobs are hard to come by in China and Thailand and these people are the best.

It is a good system used by many. The arrangement works because of trust and necessity. I think it works far better than contracts. The IRS was satisfied, but I doubt they really believed someone could trust someone else with millions of dollars and not be concerned. It is hard to believe – but also a sad commentary – about how most business is done.

Today, I spend most of my time ferreting out new opportunities in diamond growth – new countries we should be investigating in case China implodes or becomes too expensive, a real possibility as I write this. We have had it happen several times in the past and I see no reason it can't happen again. My efforts have turned to making better and bigger diamonds.

I was very impressed with New Diamond Technology (NDT) company located in St. Petersburg, Russia and was invited to their plant and given a quick overview tour. Not advisable to someone like me, but I did learn something, which I will not divulge. NDT makes the biggest, cleanest HPHT diamonds in the world, or did so up until about 2019. Something happened and they moved to Kyiv, Ukraine, not a positive move in my opinion, based on my experiences there.

Laboratory-grown diamonds manufactured by Chatham Created Gems. This image is featured in "Laboratory-Grown Diamonds," an informational guide to HPHT-grown and CVD-grown diamonds by Dusan Simic and Branko Deljanin. Colored diamonds were the easiest for Chatham to grow. It took years to unlock the secret of growing white or colorless diamonds, which is the largest market for gem diamonds. Credit: Julia Kagantsova.

They are using the Chinese multi-directional press, as are many others. The bare presses are $500,000 each and that's just a start because it does not come with instructions or the software to produce diamonds.

I have been asked many times to sell someone a diamond press and I always turn them down. Many people think you can just throw a handful of graphite in the middle chamber and flip a switch and get diamonds. I often explain that this is like buying a jet plane with no flying experience. It's a guaranteed disaster.

Many very intelligent scientists and physicists have tried to make diamond over the last 200 years and the successes can be counted on one hand. I will decline to explain the difficulties involved because they are top secret in every company, but there have been leaks. That's why the production of laboratory-grown diamonds stands at about 5 billion carats per year. Most is powder for industrial use, but more and more companies are figuring out how to grow bigger and whiter stones.

Once, a group of New York diamond cutters asked to cut a few of my diamond crystals after I gave them a presentation in 1996 after the JCK show. "It cuts just like natural diamond!" they said. "Well, what do you expect, it is diamond!" I responded. That is the hardest thing to get through people's heads: If I say this is an emerald, then it is. If I say this is identical to natural diamond, it is a diamond! People just have a hard time with chemistry, I guess. It's not magic; it's not alchemy. It's just plain inorganic chemistry.

There is a preordained order of atomic structure of the elements on the Periodic Table of elements which if "bent" or "changed" will

{217}

A Chatham lab-grown 2,134 carat ruby. One of the world's largest lab-grown rubies, this monster was grown by accident when a supporting platinum wire broke and allowed the seed plate to change the axis of growth material passing by. It totally changed how we grew all ruby and sapphires from that day on. Some accidents proved invaluable, but not many! Credit: Orasa Weldon.

result in a different compound. In the case of diamond, the atomic arrangement of carbon atoms is an absolute, and therefore will have all the physical, optical and structural qualities as its natural counterpart.

If you "play" with the structure, you will end up with a different form of carbon. Not enough pressure and you get graphite, a black oily carbon used in pencils. Add a few atoms of boron and you get blue diamonds; add nitrogen you get yellow diamonds. Identifying these trace elements and where they sit in the structure of diamond is the only way one can separate diamonds from the earth from a laboratory-grown diamond.

Is there opposition? Of course. Just like when Carroll Chatham announced his emerald success and the trade went berserk. The diamond business, one thousand times the size of the colored stone business, has gone completely over the edge in their pursuit of our demise. The rhetoric reached such a heated level that the FTC had to step in and create some guidelines in 2018 to prohibit product bashing and the release of false information. One important milestone was the declaration from the FTC that all diamonds are created equal, be they mined or lab-grown. Being "natural" does not affect their legitimacy. Both are equally diamond.

My number one endeavor today is to try and defuse this animosity and bring people together. I am a member the International Grown Diamond Association (IGDA), which has about 50 members who are producers, resellers and retailers. We meet regularly via Zoom and discuss issues such as nomenclature rules, FTC guidelines and how to properly position ourselves to the consuming public. That last issue is still under debate and will no doubt be unresolved for years to come. The natural diamond sector is scared to death of the possibility of laboratory-grown diamonds taking its place and has pulled out all stops when it comes to propaganda and fake news. The IGDA can be located at www.theigda.org where all founding members, including Chatham, have collectively stated their personal desire to better educate the consuming trade and their customers through education. Our goals include promoting ethical industry practices and standards among members to offset inaccurate information from the natural diamond sector.

Example: According to the Natural Diamond Association, "Lab grown diamonds are worthless and will soon be priced like CZ (cubic zirconium)." The simplicity of CZ makes it possible to produce it in batches by the ton that need to be lifted from reactors by crane. The equipment required to produce

HPHT or CVD diamonds, on the other hand, is very complicated and the environment needed to do so is very extreme. Also, diamond is much more complicated in its structure and can't be made by the ton, so it will never be cheap to produce diamond. This and other misleading statements are constantly coming out of the natural diamond sector.

This book will be finished long before these issues are settled. One of my main objectives today is to get the two industries talking and to stop throwing rocks at each other. It only hurts the business overall if it continues. People don't "need" diamonds – but we need people to love diamonds and want to buy them. Fighting each other is counterproductive.

I have been through this before with emeralds and rubies and it was counterproductive then, too. Once we agreed that we could ignore each other and market our individual stones without bad-mouthing the other, the industry expanded and both sectors thrived. Natural emerald and ruby are at the pinnacle of success today and so are lab-created emeralds and rubies. There are two distinct markets, both doing well.

Gemologically, there are no more holy grails to achieve. Every important stone has been created in labs around the world. Some are just curiosities because the cost to produce them versus the cost of a natural stone (such as tourmaline or tanzanite) is not viable.

The growers of lab-created diamonds are reaping the benefits of Carroll Chatham's efforts to market the gemstones he created and his company's 80 years of influence, which has left an indelible mark on the jewelry industry worldwide today.

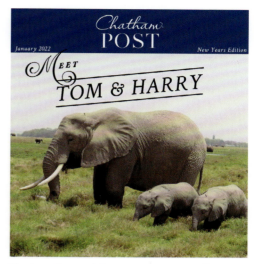

Chatham Created Gems Inc. supports efforts to end elephant poaching in Africa.

INTRODUCING TWIN BABY ELEPHANTS, TOM & HARRY!

2022 is the year of giving back! In support of the Elephant Cooperation, a non-profit organization devoted to raising awareness of the African Elephant crisis, Chatham recently participated in a live auction event. Our contribution won us a unique opportunity to name a pair of rare identical twin boy elephants! We proudly introduce you to Tom & Harry, the newest members of the Chatham family!

A plaque with their names are on display at the Amboseli National Park in Kenya. Only 1% of elephant births are twins! This is truly a magical start to a new year.

Chatham is motivated to continue spreading the word about the dangers African elephants face everyday. Poaching is a continual threat in this region endangering the African elephant population.

In addition to supporting elephants, the coalition also raises funds for local farming, child education, animal relocation and many other initiatives to help the Kenyan region thrive.

Chatham is asking for $5 or $10 donations with every loose lab grown gemstone or jewelry order. Your contribution will go directly to the Elephant Cooperation!

We strongly believe it is time for the industry to give back to a continent that has already given so much to us! Your charitable donation will help rebuild villages and protect wildlife.

If you'd like to make additional donations, we are accepting that too!

Visit www.elephantcooperation.com for more information.

"Elephant Cooperation is a unique 501(c)3 focusing on nonprofit cooperation, best business practices, and creating leaders."
EIN: 81-3209656

Afterword

It's been a wonderful life, playing chemist, jeweler, gemologist and CEO of Chatham Created Gems, Inc. over the last 55 years.

I learned a lot from my father, but he rarely actively taught me. He was quiet and introverted and always thinking. I have no idea what was going through his head when he would just stand at one of the benches in the lab, or on his porch in his modest country home 60 miles away from San Francisco in the redwoods. He never sat to think (just to eat), a habit I have picked up from him. It just seems to stimulate the thought process better than sitting.

My father was no doubt a genius in his field of inorganic chemistry. He had a gift I don't even think he realized he possessed. He taught me a lot about chemistry in areas I never would have picked up in school. Sometimes we would get into discussions of how to build something or how to evaluate some recent production occurrences, be they good or bad. We fabricated 95% of what we used to produce gemstones, so engineering problems were common.

There were no "stupid" opinions one could give my father, just arguable ideas. He taught me to think and ask questions, to never say, "I can't" do this or that. There was always a way, in his mind. One that always got my hackles up was "I already tried that" type of answer. Then off we would go, arguing the differences between what he "tried" and my idea to try again, only a little differently. There were no wrong answers; there was no pulling rank when it came to making decisions.

My father taught me to realize there is a lot more at play in growing crystals than just chemistry. He explained to me how a geode is formed in the earth and how similar that was to what we were doing. The physical aspects of what makes crystals grow and stop growing; the uniqueness of gem deposits and how to apply those physical aspects to what we were doing inside a furnace. Once you really wrapped your mind around why crystals grow in the earth's crust, you have solved most of your questions.

{220}

Duplicating those conditions was another problem altogether. He taught me to think outside the box, so to speak, when it came to the physical set up of our growth chambers, an area we still ponder today. So many little things can make or break a crystal run. Being able to understand and comprehend those little nuances was his guiding light.

I remember vividly an occasion of immense consequences in our ruby production. The physical set up inside a growth chamber is very critical and the changes that occur physically during an eight-month growth cycle are fascinating. The crystals grow bigger and divert the natural convection of flux caused by the heating element on the bottom, almost like watching water boil, before the bubbles start. Heat rises and that is convection, be it in a volcano or a furnace at 1,200 Celsius. On one run, one seed suspension wire broke, so instead of the nutrient hitting the flat side of a seed exposing the C axis of the crystal, the seed now had the A axis being bathed in nutrient.

The results were unbelievable, the largest crystals grown to date. One was over 3,000 carats!! I kept that one for our crystal museum in the corporate office in San Marcos.

Because of the way we approach each end of run inspection, we could easily see what happened after the cool down. It was extraordinarily stupid and simple. My father couldn't explain why it was better to grow on that axis, but we all agreed, it was the way to go.

These and other little "accidents" are what made it so much fun to work in the labs with Carroll Chatham. We never ceased to enjoy opening a new run.

I hope you enjoyed this walk through history – I certainly enjoyed the trip!

Acknowledgements

I want to thank the many people who have encouraged me to write this book over the last 20 years. I reached out to friends for advice on publishing and editing as I worked on it, including Dona Dirlam, who ran the Richard T. Liddicoat Gemological Library and Information Center at GIA for more than 30 years. She was very supportive of this effort and encouraged me to write it, to document our contributions to created gemstones for gem and jewelry history.

From Dona, I found Amanda Luke, also formerly from GIA, who became my editor and guide on almost every page! She kept me on track every step of the way. Without her effort, this work would be a wandering story with some unanswered questions.

The talented photographer and graphic designer Orasa Weldon brought the photos to life and perfected all the visual magic I was trying to share with you, the reader. Thank you Orasa for an excellent job well done.

In addition, I want to thank Justyne DeWilde, my granddaughter, for her excellent command of English grammar! I never knew she had it in her.

And to all the gemologists around the world I have met and spoken to. It has been my pleasure to meet so many people with a shared love for gemology and crystallography – be it grown in a laboratory or in the earth. A special thanks to Jim Shigley, Distinguished Research Fellow at GIA, for writing the introduction to this book. Jim has been contributing to gemological research for more than 40 years and is known throughout the world for his expertise.

And last but certainly not least, I want to thank my wife, Dianna, for all the trade shows she helped with and the endless hours flying coach with me around the world! Her companionship and endless support over millions of miles made all this endurable.

Further Reading: Published information on Chatham Created Gems

— — —

"50 years of Chatham Created Gems" (1988) *Jewelers' Circular-Keystone,* vol. 159, no. 9, pp. 114-116.

"AGS Collection Debuts" (2000 *Jeweler, vol) Modern. 99,* no. 6, pp.163.

Alexander, A.E. (1949) "The Synthesis of Rutile and Emerald," *Journal of Chemical Education,* vol. 26, no. 5, pp. 254-258.

Alexander, A.E. (1959) "The Chatham Ruby Makes its Bow," *Gemmologist,* vol. 28, no. 340, pp. 201–204.

Alperstein, E. (1993) "Tom Chatham: On the Cutting Edge of Creation," *In Focus Magazine*, pp. 28-31.

"Around the Institute: Richard T. Liddicoat, Jr., Mary Hanns, Peter C. Keller, Dona Dirlam, Tom Chatham, Jill M. Walker (Hobbs)" (1983-1984) *Through the Loupe: The GIA News,* pp. 3.

Bank, H. (1977) "Mit dem schmelzdiffusionsverfahren hergestellte synthetische korunde (rubine und saphire nach Chatham)," *Zeitschrift der Deutschen Gemmologischen Gesellschaft,* vol. 26, no. 3, pp. 170-172.

Bates, Rob (1999) "Gemology's Outer Limits," *Jewelers' Circular-Keystone,* vol. 170, no. 6, pp. 126-128.

Beres, Glen A. (2005) "Scoop on Synthetics," *Jewellery Business,* vol. 1, no. 2, pp. 38-41.

Berk, M. (1987) "Wearable Science," *Lapidary Journal,* vol. 40, no. 12, pp. 25-31.

Bosshart, George (1981) "Die Unterscheidung von echten und synthetischen Rubinen mit UV-Spektralphotometrie," *Zeitschrift der Deutschen Gemmologischen Gesellschaft,* vol. 30, no. 3/4, pp. 157-169.

Bosshart, George (1982) "Distinction of Natural and Synthetic Rubies by Ultraviolet Spectrophotometry," *Journal of Gemmology,* vol. 18, no. 2, pp. 145-160.

Breeding, Christopher; Shigley, James E.; Shen, Andy (2005) "As-grown, Green Synthetic Diamonds," *Journal of Gemmology,* vol. 29, no. 7/8, p. 387-394.

Brown, G. (1984) "Inclusions in Synthetic Corundum by Chatham," *Australian Gemmologist,* vol. 15, no. 5, pp. 149-154.

Burch, C.R. (1987) "Metallic Inclusions in Chatham Synthetic Corundums," *Journal of Gemmology,* vol. 20, no. 5, pp. 267-269.

"Chatham Celebrates Creative Chemistry," (1988) *Modern Jeweler,* vol. 87, no. 7, pp. 50, 54, 56.

"Chatham Makes Donation to Vision 2000," (1996) *Modern Jeweler*, v. 95, n. 4, p. 101.

Chatham, Carroll F. (1958) "Cultured Emeralds and How They Grow," *American Jewelry Manufacturer*, (May), pp. 18, 20, 22.

Chatham, Carroll F. (1982) "Little Known Facts in the Art of Growing Gem Cystals," *International Gemological Symposium – Proceedings*, pp. 153-156.

Chatham, Tom (1982) "Marketing of Created Gemstones," *International Gemological Symposium – Proceedings*, pp. 159-163.

Chatham, Tom (1998) "Created Gemstones, Past, Present, and Future," *Canadian Gemmologist*, vol. 19, no. 1, pp. 8-12.

Crowningshield, G. Robert (1958) "Highlights at the Gem Trade Lab in New York: Synthetic Emerald Testing," *Gems & Gemology*, vol. 9, no. 8, p. 228.

Crowningshield, G. Robert (1966) "Chatham Synthetic Ruby," *Gems & Gemology*, vol. 12, no. 4, pp. 110-112.

"Cultured War: Trade Wants C-word Terminated," (2007) *JCK*, vol. 178, no. 3, pp. 52, 54.

Cumo, Carlo (2019) "From the Analyst's Notebook: An Investigation on a Red Faceted Stone," *Italian Gemological Review*, no. 7, pp. 7-13.

Deljanin, Branko; Peretti, Adolf; Alessandri, Matthias (2019) "Origin and Grading of Pink Diamonds: Argyle and non-Argyle Natural, Treated and Lab-grown,"

Gemmologie: Zeitschrift der Deutschen Gemmologischen Gesellschaft, vol. 68, no. 1/2, pp. 26-27.

"Diamond Creators: The New Alchemists," (2004) *Israel Diamonds*, no. 193, pp. 46, 48-49.

"Diamond Update ... GIA Examines Latest Production Samples," (1997) *Chatham News Network*, n. 19.

Duyk, F. (1963) "New-type Inclusions in Chatham Synthetic Emeralds," *Journal of Gemmology*, vol. 9, no. 4, pp. 130-131.

Federman, David (1996) "After the Crash," *Modern Jeweler*, vol. 95, no. 11, pp. 20, 24, 26, 28-30, 32.

Fenelle, Cheryl (1998) "Laboratory-created Gems Grow Acceptance at Retail," *National Jeweler*, vol. 42, no. 14, pp. 74, 76, 78, 80, 82.

Flanigen, E.M.; Breck, D.W.; Mumbach, N.R.; Taylor, A.M. (1967) "Characteristics of Synthetic Emeralds," *American Mineralogist*, vol. 52, no. 5/6, pp. 744-772.

Frederick, Larry (1999) "Diamond Notes: Lab-grown 'Whites' Still Years Off," *Jewelers' Circular-Keystone*, vol. 170, no. 4, p. 50.

"GEM-A conference 2004," (2004) *Gem & Jewellery News*, vol. 14, no. 4, p. 77-79.

"Gem Notes: Chatham's Synthetic Project Moves to the States," (1999) *Jewelers' Circular-Keystone*, vol. 170, no. 2, p. 26.

"Gem Trade Lab Notes: Synthetic Ruby," (1981) *Gems & Gemology*, vol. 17, no. 3, pp. 163-165.

"GIA Inducts 1997 League of Honor Sovereigns," (1998-1999) *In Focus*, p. 20.

"GIA's 1997 League of Honor Sovereigns," (1998) *Jewelers' Circular-Keystone,* vol. 169, no. 1, p. 292.

"The Gilson Synthetic Emerald," (1970) *Swiss Watch & Jewelry Journal,* no. 5, pp. 593-595.

Gomelsky, Victoria (2001) "Growing Pains," *National Jeweler*, vol. 45, no. 14, pp. 36, 38.

Gübelin, Eduard J. (1964) "Two New Synthetic Emeralds," *Gems & Gemology*, vol. 11, no. 5, pp. 139–148.

Gübelin, Eduard J. (1982) "Erkennungsmerkmale der neuen synthetischen Saphire," *GZ: Goldschmiede Zeitung,* vol. 80, no. 11, pp. 51-57.

Gübelin, Eduard J. (1983) "Identification of the New Synthetic and Treated Sapphires," *Journal of Gemmology,* vol. 18, no. 8, pp. 677-705.

Gübelin, Eduard J. (1983) "The Recognition of the New Synthetic Rubies," *Journal of Gemmology,* vol. 18, no. 6, pp. 477-499.

Gunawardene, M. (1983) "Über Die Synthetischen Blauen und Orangefarbenen Sapphire von Chatham," *Zeitschrift der Deutschen Gemmologischen Gesellschaft,* vol. 32, no. 4, pp. 196-203.

Gunawardene, M. (1985) "Identification Characteristics of Flux Grown Synthetic Orange Sapphires," *Journal of Gemmology,* vol. 19, no. 5, pp. 389-403.

Hänni, Henry A. (1982) "A Contribution to the Separability of Natural and Synthetic Emeralds," *Journal of Gemmology,* vol. 18, no. 2, pp. 138-144.

Hänni, Henry A.; Kiefert, Lore (1994) "AGEE Hydrothermal Synthetic Emeralds," *JewelSiam*, vol. 5, no. 5, pp. 80-85.

Hodgkinson, A. (1995) "The Hanneman-Hodkinson Synthetic Emerald Filter," *Canadian Gemmologist,* vol. 16, no. 1, pp. 18-22.

Huong, L.T.T.; Hofmeister, W., Häger T.; Karampelas, S.; Kien, N.D.T. (2014) "A Preliminary Study on the Separation of Natural and Synthetic Emeralds Using Vibrational Spectroscopy," *Gems & Gemology,* vol. 50, no. 4, pp. 287-292.

Johnson, P.W. (1961) "All About Emeralds Natural or Synthetic," *Lapidary Journal,* vol. 15, no. 1, pp. 118-131.

Kammerling, Robert C.; Koivula, John I.; Fritsch, Emmanuel. (1994) "An Examination of Chatham Flux-grown Synthetic Pink Sapphires," *Journal of Gemmology,* vol. 24, no. 3, pp. 149-154.

Kane, Robert E. (1982) "The Gemological Properties of Chatham Flux-grown Synthetic Orange Sapphire and Synthetic Blue Sapphire," *Gems & Gemology,* vol. 18, no. 3, pp. 140-153.

Kiefert, Lore; Schmetzer, Karl. (1988) "Morphology and Twinning in Chatham Synthetic Blue Sapphire," *Journal of Gemmology,* vol. 21, no. 1, pp. 16-22.

Kiefert, Lore; Schmetzer, Karl (1992) "The Microscopial Determination of Structural Features of Uniaxial Natural and Synthetic

Gemstones Part III," *Antwerp Gems,* vol. 3, no. 2, p. 25-13.

Kiefert, Lore; Schmetzer, Karl (1988) "Morphology and Twinning in Chatham Synthetic Blue Sapphire," *Journal of Gemmology,* vol. 21, no. 1, p. 16-22.

Koivula, John I. (1980) "Gemological Notes: Brief Notes on Chatham Flux Sapphires," *Gems & Gemology,* vol. 16, no. 12, pp. 410-411.

Koivula, John I.; Fritsch Emmanuel (1994) "Chatham Flux-grown Pink Synthetic Sapphires," *Gems & Gemology,* vol. 30, no. 1, pp. 56-57.

Koivula, John I.; Kammerling, Robert C. (1989) "Chatham Expands," *Gems & Gemology,* vol. 25, no. 2, pp. 117-118.

Koivula, John I.; Kammerling, Robert C. (1991) "Update on Chatham Production," *Gems & Gemology,* vol. 27, no. 1, pp. 53-54.

Lens, L. (1987) "The Development of the Lennix Synthetic Emerald," *Gem Instrument Digest,* vol. 3, no. 2, pp. 26-28.

Liddicoat, Richard T. (1970) "Hexagonal Platelets in Chatham Flux-melt Rubies," *Gems & Gemology,* vol. 13, no. 7, pp. 234-235.

Liddicoat, Richard T. (1983) "Carroll Chatham: 1914-1983," *Gems & Gemology,* vol. 19, no. 3, p. 129.

Lijian, Qi; Yuan, Joe C. C.; Haibin, Liu (1998) "Characteristics and Identification of Chatham Synthetic Diamond," *China Gems,* vol. 7, no. 2, pp. 55-59.

Ljian, Qi; Xinqiang, Yuan; Yongan, Luo; Zhizhong, Yuan (1999) "Chatham Synthetic Colourless Diamond – A Mixed Variety of IIa ~ IIb," *Journal of Gems & Gemmology,* vol. 1, no. 4, pp. 7-10.

Montgomery, Lee; Roderick, Kyle (1992) "All in the Family," *In Focus,* vol. 11, no. 2, pp. 10-14, 30-35.

Nassau, Kurt (1976) "Synthetic Emerald: The Confusing History and the Current Technologies," *Journal of Crystal Growth,* vol. 35, no. 2, pp. 211-222.

Nassau, Kurt (1980) *Gems Made by Man,* Chilton Book Company, Radnor, Pennsylvania, 364 pp.

Nassau, Kurt (1980) "Synthetics in the Seventies," *Lapidary Journal,* vol. 34, no. 1, pp. 50-51, 54,56,58,64, 66-68.

Nassau, Kurt (1982) "Colored Synthetics and Imitations," *International Gemological Symposium – Proceedings,* pp. 143-150.

Nassau, Kurt (1987) "The Current Decade - Synthetic Gemstones in the 1980s," *Lapidary Journal,* vol. 40, no. 12, pp. 32-33, 35-42.

Nassau, Kurt (1990) "Synthetic Gem Materials in the 1980s," *Gems & Gemology,* vol. 26, no. 1, pp. 50-63.

Nassau, Kurt, Nassau, J. (1980) "The Growth of Synthetic and Imitation Gems," *Crystals: Growth, Properties and Applications,* Vol. 2, pp. 1-50.

"News flash from ADAMAS Gemological Laboratory: Near Colourless Synthetic Diamond Diagnostic Feature," (2001)

Gemmology Queensland, vol. 2, no. 5, p. 22-23.

O'Donoghue, M. (1983) "Orange Synthetic Corundum," *Journal of Gemmology,* vol. 18, no. 8, pp. 736-737.

O'Donoghue, M. (2005) *Artificial Gemstones,* N.A.G. Press, London, 294 pp.

Ponalho, J. (1988) "Quantitative Cathodoluminescence - A modern Approach to Gemstone Recognition," *Journal of Gemmology,* vol. 21, no. 3, pp. 182-193.

Pough, F.H. (1963) "The Synthetic Emerald Family," *Lapidary Journal,* vol. 17, no. 3, pp. 380-387.

Qi,L.; Xi, J.; Pei, J. (1999) "Chatham Flux-grown Synthetic Ruby: A New Variety with Dark Colored Cores," *Journal of Gems and Gemmology,* vol. 1, pp. 1-8.

Read, Peter (1988) "A Fair Selection," *Canadian Jeweller,* vol. 109, no. 3, p. 12.

"Real, cultured or cloned?" (2004) *SolitairePro,* pp. 38-41.

Reinecke, C. (1959) "The New Rubies of Carroll Chatham (with a discussion of rubies in general)," *Lapidary Journal,* vol. 13, no. 4, pp. 528-530.

"Revitalized AGTA Sets Sights on Gem Promotion," (1997) *Jewelers' Circular-Keystone,* vol. 168, no. 8, p. 46.

Rogers, A.F.; Sperisen, F.J. (1942) "American Synthetic Emerald," *American Mineralogist,* vol. 27, no. 11, pp. 762-768.

Roskin, Gary (2004) "Jewel of the Month: Gem Quality Synthetic Diamond," *Jewelers' Circular-Keystone,* vol. 175, no. 2, pp. 72-74.

Saesaew, Sudarat (2020) "Lab Notes: Unusual Absorption in a Blue Flux-grown Synthetic Sapphire," *Gems & Gemology,* vol. 56, no. 4, pp. 524-525.

Scarratt, Kenneth (1977) "A Study of Recent Chatham Synthetic Ruby and Synthetic Blue Sapphire Crystals with a View to the Identification of Possible Faceted Material," *Journal of Gemmology,* vol. 15, no. 7, pp. 347-353.

Scarratt, K.; Dunaigre, C.M.; DuToit, G. (1996) "Chatham Synthetic Diamonds," *Journal of the Gemmological Association of Hong Kong,* vol. 19, pp. 6-12.

Schmetzer, Karl (1986) "Production Techniques of Commercially Available Gem Rubies," *Australian Gemmologist,* vol. 16, no. 3, pp. 95-100.

Schmetzer, Karl; Kiefert, Lore; Bernhardt, Heinz-Jürgen (1999) "Multicomponent Inclusions in Nacken Synthetic Emeralds," *The Journal of Gemmology,* vol. 26, no. 8, pp. 487-500.

Shigley, James E. (2004) "Synthetic Coloured Diamonds Sold by Chatham Created Gems," *Indian Gemmologist,* vol. 12, no. 1, pp. 7, 14-15.

Shigley, James E.; McClure, Shane F.; Breeding, Christopher M.; Hsi-Tien, Andy; Muhlmeister, Sam M. (2004) "Lab-Grown Colored Diamonds from Chatham Created Gems," *Gems & Gemology,* vol. 40, no. 2, pp. 128-145.

Shigley, James E.; Breeding, Christo-

pher; Shen, Andy (2005) "New Laboratory-Grown diamonds from Chatham Created Gems," *Rapaport Diamond Report*, vol. 28, no. 9, pp. 46-49.

Solotaroff, Ivan (2008) "Counter Cultured," *Modern Jeweler*, vol. 107, no. 5, pp. 58-60, 62-66.

Stockton, Carol M. (1984) "The Chemical Distinction of Natural from Synthetic Emeralds," *Gems & Gemology*, vol. 20, no. 3, pp. 141-145.

Stockton, Carol M. (1987) "The Separation of Matural from Synthetic Emeralds by Infrared Spectroscopy," *Gems & Gemology*, vol. 23, no. 2, pp. 96-99.

Sunagawa, I. (1964) "A Distinction Between Natural and Synthetic Emeralds," *American Mineralogist*, vol. 49, no. 5/6, pp. 785-793.

Switzer, G.S. (1946) "Improvements in Quality in Synthetic Emerald," *Gems & Gemology*, vol. 5, no. 5, pp. 305-307.

Thompson, Sharon Elaine (2008) "Smokin' Stones: Lab-grown Diamonds," *Lapidary Journal Jewelry Artist*, vol. 61, no. 12, p. 18.

Wakefield, Sharon (1997) "Synthetic Diamonds: Dealing with New Realities," *Rapaport Diamond Report*, vol. 20, no. 22, pp. 67, 71.

Weldon, Robert (2003) "Color Made to Order," *Professional Jeweler*, vol. 6, no. 9, pp. 30, 32-33.

Weldon, Robert (1998) "Growing Carats," *Professional Jeweler*, vol. 1, no. 1, pp. 47-48.

Whittaker, W. (1951) "They Make Jewels at Home," *Popular Mechanics*, vol. 96, no. 3, pp. 89-93 and 228 and 234 and 236.

Yonick, Deborah (1999) "Chatham's New Creation," *Professional Jeweler*, vol. 2, no. 8, p. 43.

Yu, K. N.; Tang, S. M.; Tay, T. S. (2000) "PIXE Studies of Emeralds," *X-RAY SPECTROMETRY*, vol. 29, pp. 267-278.

Zeitner, J.C. (1987) "No 'Sin' in Synthetics," *Lapidary Journal*, vol. 40, no. 12, pp. 20-24.

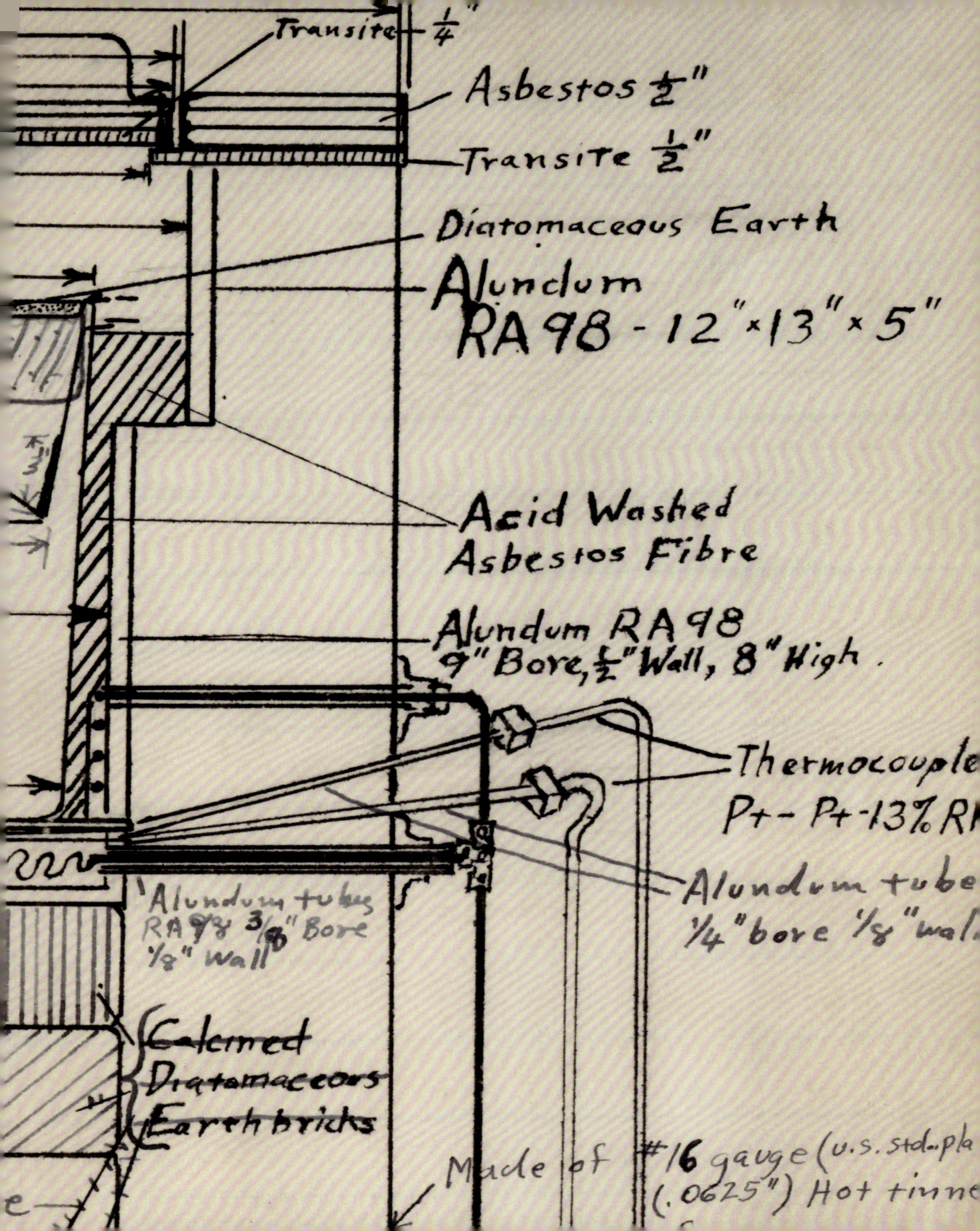

Transite $\frac{1}{4}$"

Asbestos $\frac{1}{2}$"

Transite $\frac{1}{2}$"

Diatomaceous Earth

Alundum
RA 98 - 12" × 13" × 5"

Acid Washed
Asbestos Fibre

Alundum RA 98
9" Bore, $\frac{1}{2}$" Wall, 8" High.

Thermocouple
Pt - Pt - 13% Rh

Alundum tube
$\frac{1}{4}$" bore $\frac{1}{8}$" wall.

Alundum tubes
RA 98 $\frac{3}{8}$" Bore
$\frac{1}{8}$" Wall

Calcined
Diatomaceous
Earth bricks

Made of #16 gauge (U.S. std. pla
(.0625") Hot tinne

About the Author

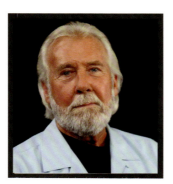

Thomas H. Chatham grew up in San Francisco with a front row seat to his father Carroll Chatham's pioneering crystal growth work. He learned early on from his father that if you wanted something, you needed to get a job to buy it. He sold Christmas wreaths, had a paper route, and bought his first car at 15 – even though he didn't have a driver's license yet. Tom wanted to love chemistry as much as his father did, but he was drawn more to making things, such as the car engines he tinkered with at an after-school job at an auto body/engine repair shop.

He spent a little more than two years in college in Santa Cruz, CA before joining his father at Chatham Research Labs in 1965. He knew he could learn more working at his father's side than he would in school and soon knew how to build furnaces, document experiments, and refine the growing process and materials used. He was involved in the first flux ruby research, followed by the development of many sapphire colors and a flux-grown alexandrite.

Tom was instrumental in establishing business relationships with international production facilities to cut their rough gemstones and then helped his father figure out how to market them to the gem and jewelry industry and jewelry loving consumers.

He joined the American Gem Society (AGS) in 1970 and received his Graduate Gemologist diploma from the Gemological Institute of America some years later to become a registered supplier of AGS. He has been a featured speaker at AGS Conclaves and spoke to GIA students about lab-grown gemstones for more than 30 years (every 6 months) hoping to teach every student about laboratory-grown gemstones. Tom spoke to approximately 20,000 students and each received a Chatham emerald crystal.

In addition, Tom spoke to many gemological groups worldwide, including Gem A in Great Britain (formally known as the Gemmological Association of Great Britain), the Gemmological Association of Hong Kong, GIA Japan, the Brazilian Institute of Gems and Precious Metals, the Canadian Gemmological Association, the Gemological Institute of Taiwan, and numerous universities, including Stanford University's material research division. Tom never hesitated to spread his knowledge of laboratory-grown gemstones.

Carroll Chatham was not able to fulfill his boyhood dream of making diamond, so Tom pursued it for him after his death. This took him on challenging, worldwide adventures until 1993, when Chatham Created Gems Inc. announced its introduction of laboratory-grown diamonds from Siberia. The search for a more beautiful laboratory-created diamond expanded to Ukraine, and finally, he introduced colorless lab-grown diamonds at the 1996 JCK Show in Las Vegas. The Chatham company next introduced lab-grow pink, yellow and blue diamonds in 2001.

Tom has authored numerous articles and editorials for the major jewelry trade publications over the years, such as *National Jeweler*, *Jewelers Circular Keystone (JCK)*, *Modern Jeweler*, *The Goldsmith*, *Professional Jeweler* and a number of international magazines. He is a founding member of the International Grown Diamond Association (IGDA), which works to educate consumers on lab-grown diamonds and promote ethical industry standards for lab-grown diamonds.

Tom stepped down as president of Chatham Created Gems in 2012, becoming the CEO until 2018 and remains Chairman of the Board. He and his wife Dianna live in the San Francisco Bay area and often enjoy month-long vacations in Cabo, Mexico. After years of worldwide travel, Mexico is about as far as he wants to go today.

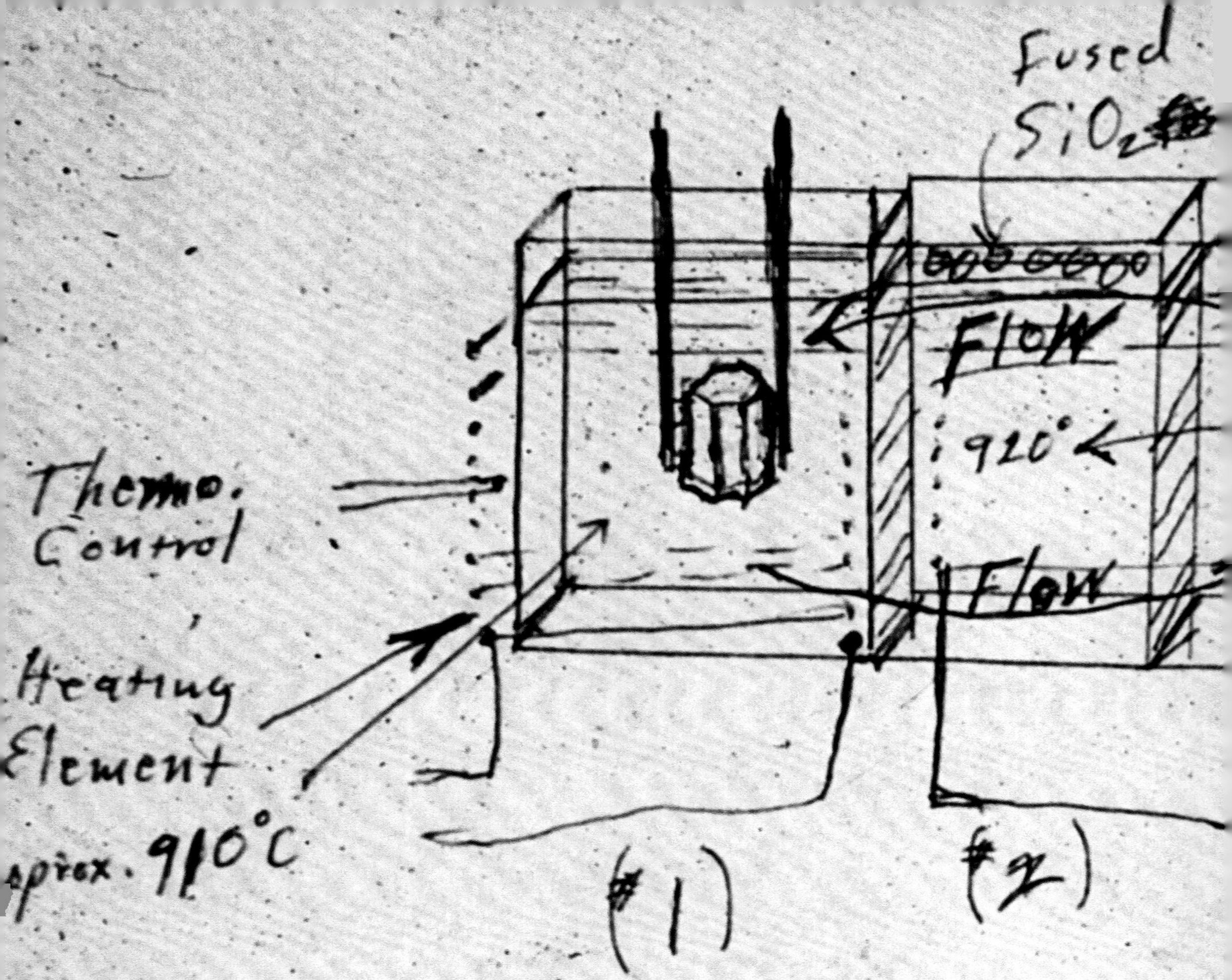

Fused SiO_2

Thermo.
Control

Heating
Element
aprox. 910°C

FLOW
920
Flow

(#1)   (#2)

. With the temp. of the Sln. (.

in the Growth chamber at #1 or

chamber #3 the use of separat

both the ~~flow of a~~ rate of Flo

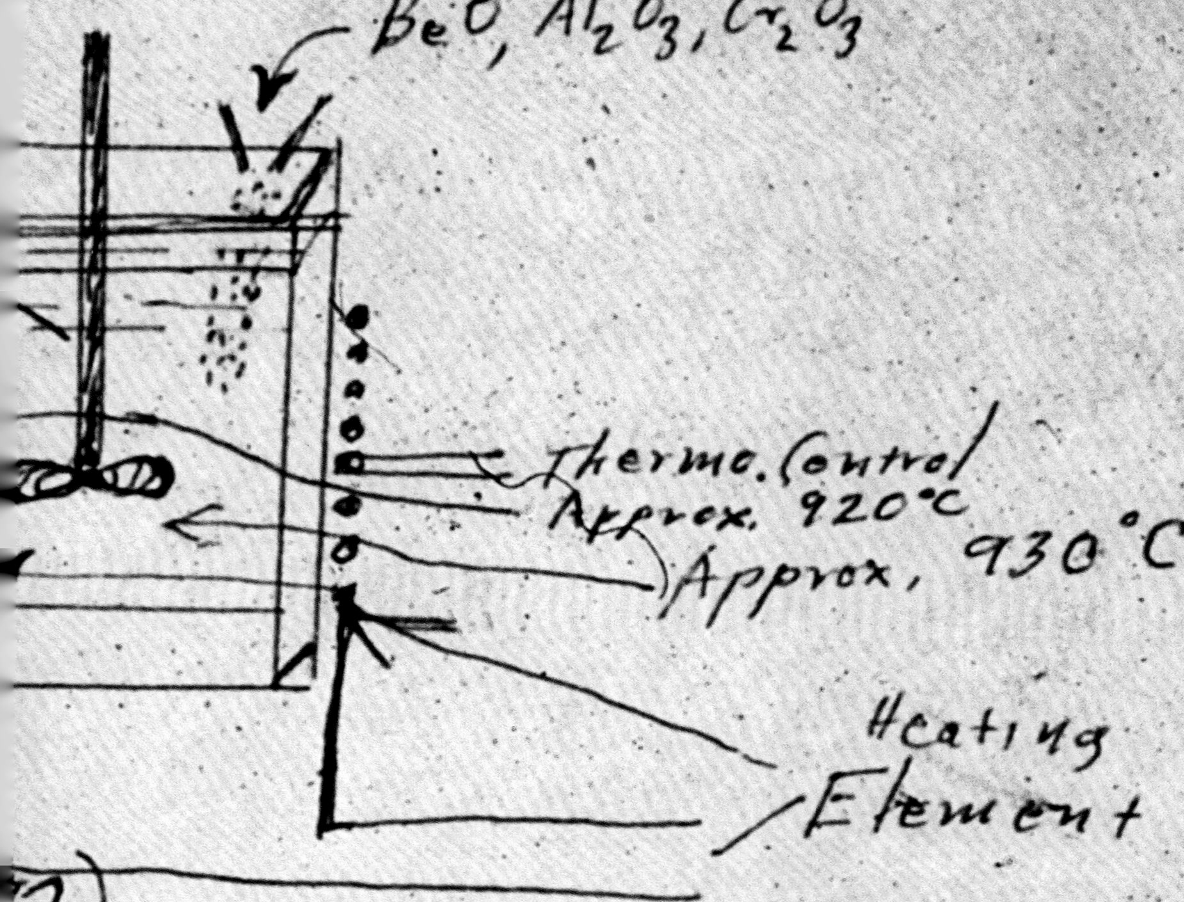

BeO, Al_2O_3, Cr_2O_3

Thermo. Control
Approx. 920°C
Approx. 930°C

Heating
Element

3)

$Mo_2O_7 \cdot V_2O_5$) Controlled at 910°C.

d 930°C. in the oxide feed

heating elements is necessary.

of Sln. and the temp difference